KinderFolders for Math Readiness

by Lillian Lieberman
illustrated by Marilynn G. Barr

Publisher: Roberta Suid
Copy Editor: Carol Whiteley
Production: Marilynn G. Barr

For a complete catalog, please write to the address below:
P.O. Box 1680, Palo Alto, CA 94302

ISBN 1-878279-82-3

Printed in the United States of America

987654321

TABLE OF CONTENTS

General Directions 5
1 Musical Shape-ups (Identifying shapes) 9
2. Shape Search (Shape pattern sequence) 13
3. Pack Rat Counting (Counting to 20) 17
4. Turtle and Hare (Counting to 50) 21
5. Tracking with Cars (Left-right directional skills) 25
6. Mr. and Mrs. Kangaroo's Pockets (Classification) 29
7. Big Fish, Little Fish (Size match—big/small) 33
8. Boxes and Balls (Size sequence—big/small) 37
9. Snakes in the Grass (Length—short/long) 41
10. Duck and Bear Look-alikes (Concept of sameness) 45
11. Bones (Number pattern sequence) 49
12. Mittens (Counting/1-1 matching) 53
13. Very Beary T-shirts (Identifying number symbols and words) 57
14. Mother Wolf's Quilts (Counting and number sets) 61
15. Space Walk (Counting/matching number symbols to sets) 65
16. Balloons (Number sequence/before and after) 69
17. Ice Cream Cones (Number sequence/number in between) 73
18. Tricky Clowns (Beginning graphing) 77
19. Grasshopper's Pogo Hop (Counting forward on number line, 0-20) 81
20. Mole's Stroll (Counting backward on number line, 20-0) 85
21. Ladybugs (Which is more?) 89
22. Apples and Worms (Which is less?) 93
23. Muffins (Recognizing the + sign and the = sign) 97
24. Nuts and Squirrels (Recognizing the - sign and the = sign) 101
25. Rabbit Families (Different names for sets) 105
26. Pumpkin Patch (Subtraction number sequence with symbols) 109
27. Card Mates (Matching the number of things to the symbols) 113
28. Coats (Object and number closeups/number writing) 117
29. Dancing Stars (Adding 1 or 2 to a whole number) 121
30. Ships Ahoy! (Taking away 1 or 2 from a whole number) 125
31. Spaceships (Counting by 1's, 2's, 5's, and 10's) 129
32. Cat, Dog, and Mouse Race (Beginning probability) 133
33. Pies (Beginning fractions) 137
34. Mouse Time (Telling time) 141

Introduction

KinderFolders for Math Readiness presents activities to enrich readiness concepts and number skills for preschoolers and first graders in easy-to-make-and-use file folder set-ups. The folders can be used with individual children, small cooperative groups, or in learning centers.

The activities in *KinderFolders for Math Readiness* help to reinforce and motivate math learning in an enjoyable format. To develop pre-academic readiness skills, children do such work as match basic shapes to musical characters, track left to right with cards, match little fish to big fish, and match bear and duck look-alikes. Number concepts are fostered by such activities as matching symbols to objects, matching sets to number symbols, and playing number dominoes. Activities that focus on beginning subtraction and addition skills and the concept of more or less are also included. A variety of hands-on responses, including placing objects, clipping on clothespins, turning a spinner, and making yarn matches, keep the children actively engaged.

Titles of related books are included to make the literature connection. These books may be used to help initiate the activities as well as to enhance the children's enjoyment and involvement in learning. The books should be read to the children and made available for their use.

Each instant folder includes the file folder layout and the activity to be duplicated, with simple directions for use. A tab label and an illustration for the folder cover are also provided. General construction and use directions follow here. The "How to Make Instant Folders" section provides additional information for particular activities.

General Directions

Construction

Use sturdy colored file folders for the instant folders. Duplicate the inside file folder set-up, the illustration for the activity, and the activity label. Color with felt pens, colored pencils, or crayons. Then trim and cut out. Glue the file folder set-up to the inside of the folder and the illustration to the outside front. Glue the directions with the "Book to Read" reference below the illustration. Glue the tab label onto the file folder tab. Laminate both sides of the file folder.

Glue the duplicated sheets, along with any loose parts, such as markers or spinners, onto oak tag for sturdiness. Color and laminate. Cut out or trim as necessary and complete construction. A craft knife is recommended for making the slits and slots for the activities that require them. Buttons or other objects can be used for markers that are not provided. Keep all loose parts in a 7 1/2" by 10 1/2" clasp envelope glued to the back of the folder.

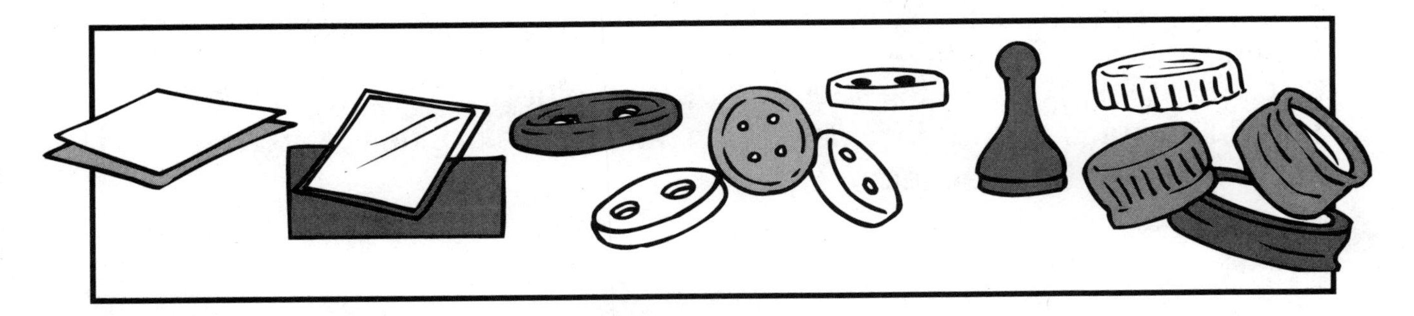

Activity Use

Have the children take out the activity and open the file folder on the work area. Instruct the children on how to play, referring to the directions. Have the children place the loose parts in the envelope after play, and deposit the folder in a file folder basket. If possible, read the recommended books prior to each activity for added motivation.

How to Make Instant Folders

Match-ups and Game Boards

(Activities 1-21)

Follow the general directions. For "Pack Rat Counting," provide a die and two markers. For "Mr. and Mrs. Kangaroo's Pockets," glue two category pictures to each kangaroo; glue pictures on three sides only, leaving the top open for slip-ins. For "Ice Cream Cones," cut out the extra activity card, glue to oak tag, color, and laminate. Enclose in the envelope with a washable felt pen or erasable crayon. For "Grasshopper's Pogo Hop" and "Mole's Stroll," duplicate the number cards twice.

Clothespin Matches

(Activities 22-26)

Follow the general directions. Cut out and glue the labels or objects securely to clothespins. Reinforce with clear sealing tape if desired. You will need 10 clothespins for each activity except "Pumpkin Patch," which requires 18 clothespins. Cut out the loose parts and glue them to the designated dotted-line areas on the file folder set-up. Leave free of glue the areas where clothespins will be attached.

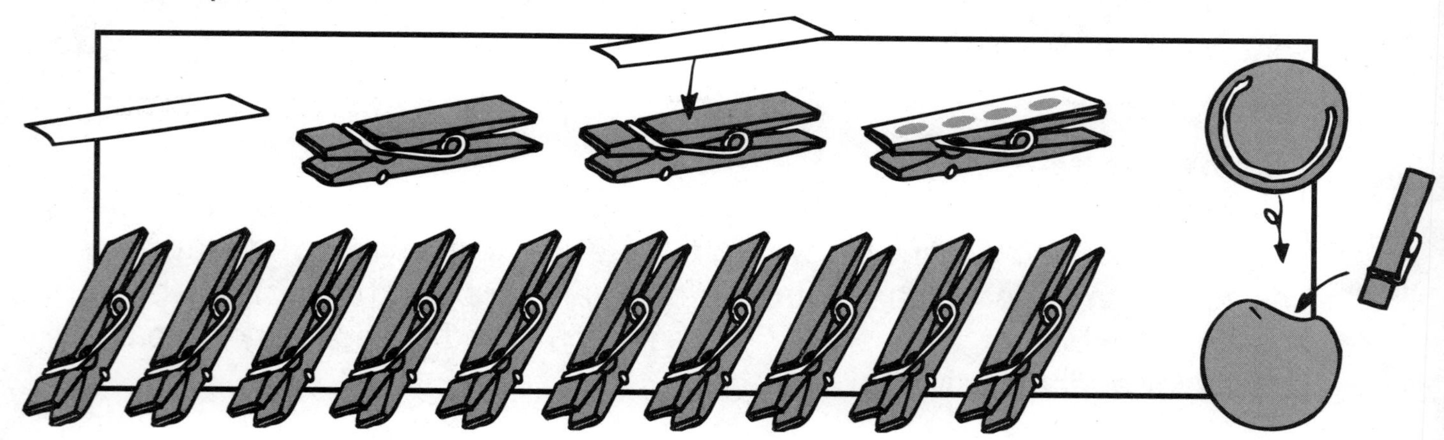

Yarn Set-ups

(Activities 27-28)

Follow the general directions. Punch holes for the yarn. Knot yarn and pull the free ends through the holes on the left side. Place brass fasteners through the holes on the right side, heads up. Glue the activity illustration and the directions to the front of the folder and laminate. For extra activity sheets, color and glue to oak tag and laminate. Enclose in the manila envelope with a washable felt pen or erasable crayon.

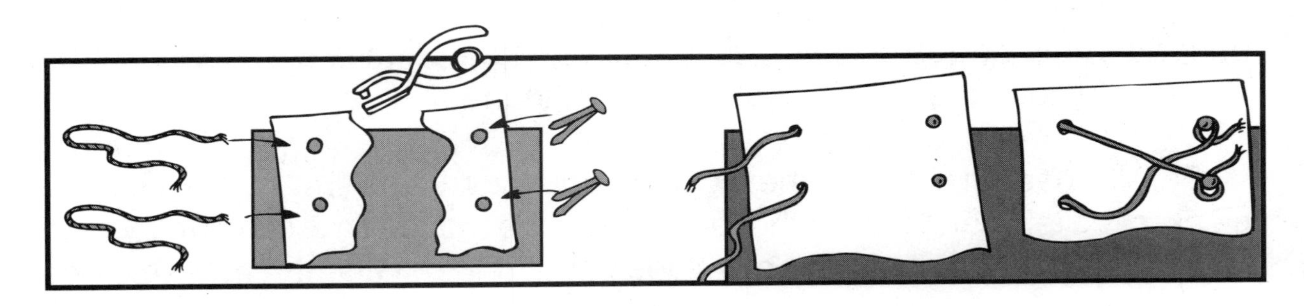

Slider Set-ups

(Activities 29-31)

Follow the general directions. Duplicate stars, ships, spaceships, and number strips. Color and glue to oak tag. Laminate. Using a sharp craft knife, cut the slits indicated on the stars, ships, and spaceships. Cut number strips out. For "Dancing Stars" and "Ships Ahoy!", slip the number strip sliders through each star and ship so that the number fact and matching answer show through the windows. Glue stars and ships to the folder, leaving the slider area free. For "Spaceships," slip the slider through the spaceship so that one number shows each time the slip is moved up or down. Glue spaceships to the file folders, leaving the slider areas free.

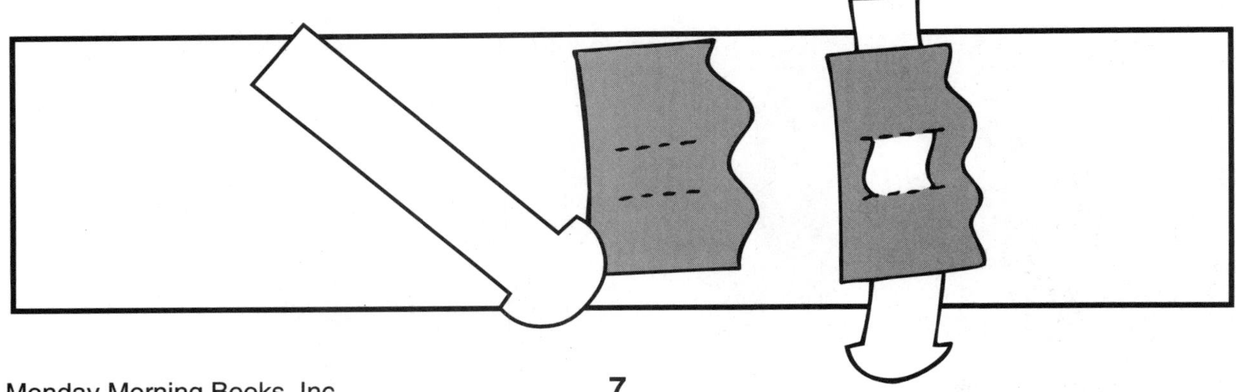

Spinner Set-ups
(Activities 32-34)

Follow the general directions. Duplicate parts for the spinner and wheel. Color and glue to oak tag. Laminate and cut out. Punch a hole in the center of the wheel with a sharp tool and attach the spinner with a brass fastener. For "Cat, Dog, and Mouse Race," color and attach the wheel and spinner to the folder. Make two copies of the tally cards. Laminate. Enclose in the manila envelope with two washable felt pens or erasable crayons.

For "Pies," make two copies of the blank pies. Color and cut out. Glue two pies to each side of the folder. Laminate folder. Make two copies of the decorated-pie pages. Color and glue to oak tag. Laminate and cut into pie pieces. Enclose the pieces with the wheel in the manila envelope.

For "Mouse Time," color the big clock and the mouse time cards. Glue the clock to the left-hand side of the folder. Glue the mouse time cards to the right side of the folder. Laminate both sides of the folder. Glue the mouse cards, mouse spinner, and cheese to oak tag. Laminate and cut out. Punch a hole in the clock and spinner and attach with a brass fastener. Enclose all the pieces in the manila envelope.

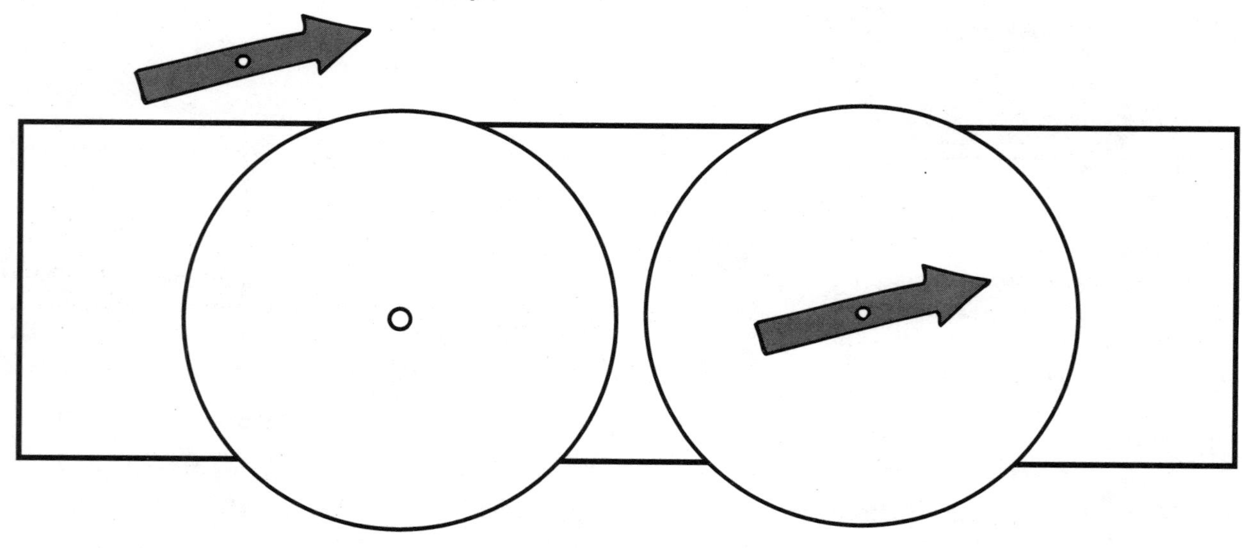

1. Musical Shape-ups

Musical Shape-ups

Identifying shapes

Directions: Take out the shape cutouts and open the folder. Name the different cutouts (oval, triangle, rectangle, circle). Put a matching shape on each musical shape in the folder. You can use the shapes to make your own shape pictures by tracing them onto a piece of paper.

A book to read: Shapes by Tana Hoban

Musical Shape-ups

Musical Shape-ups

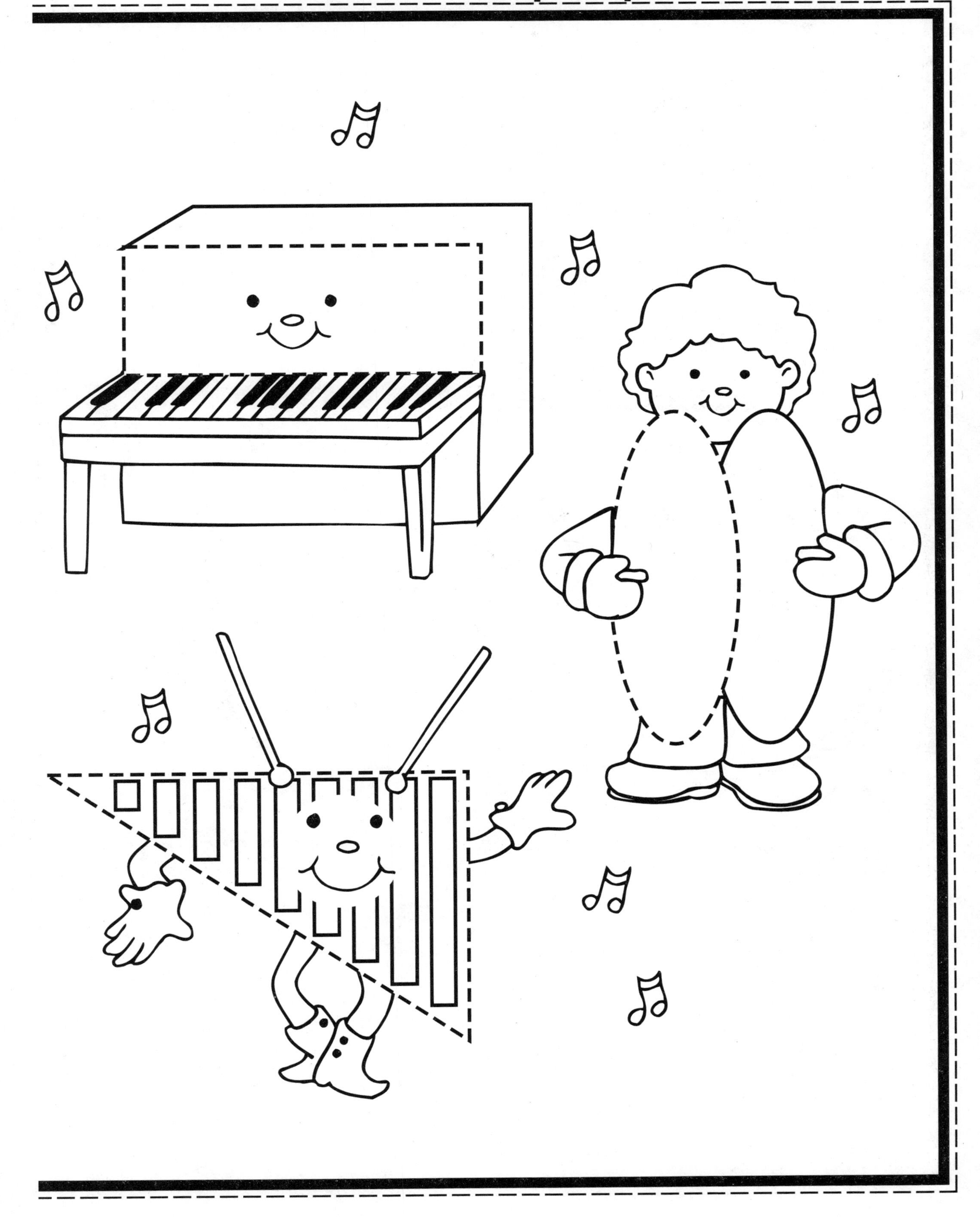

Musical Shape-ups

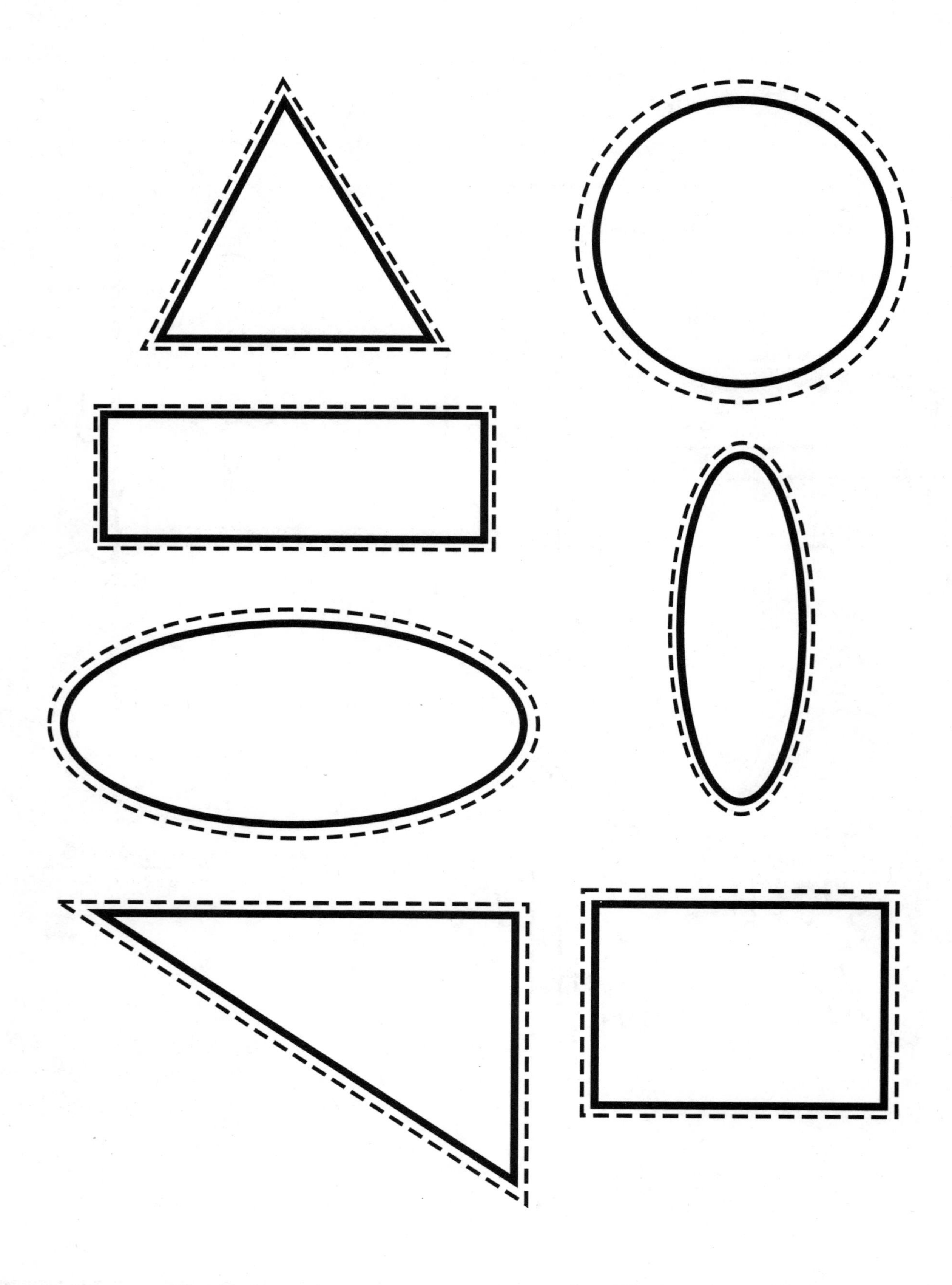

2. Shape Search

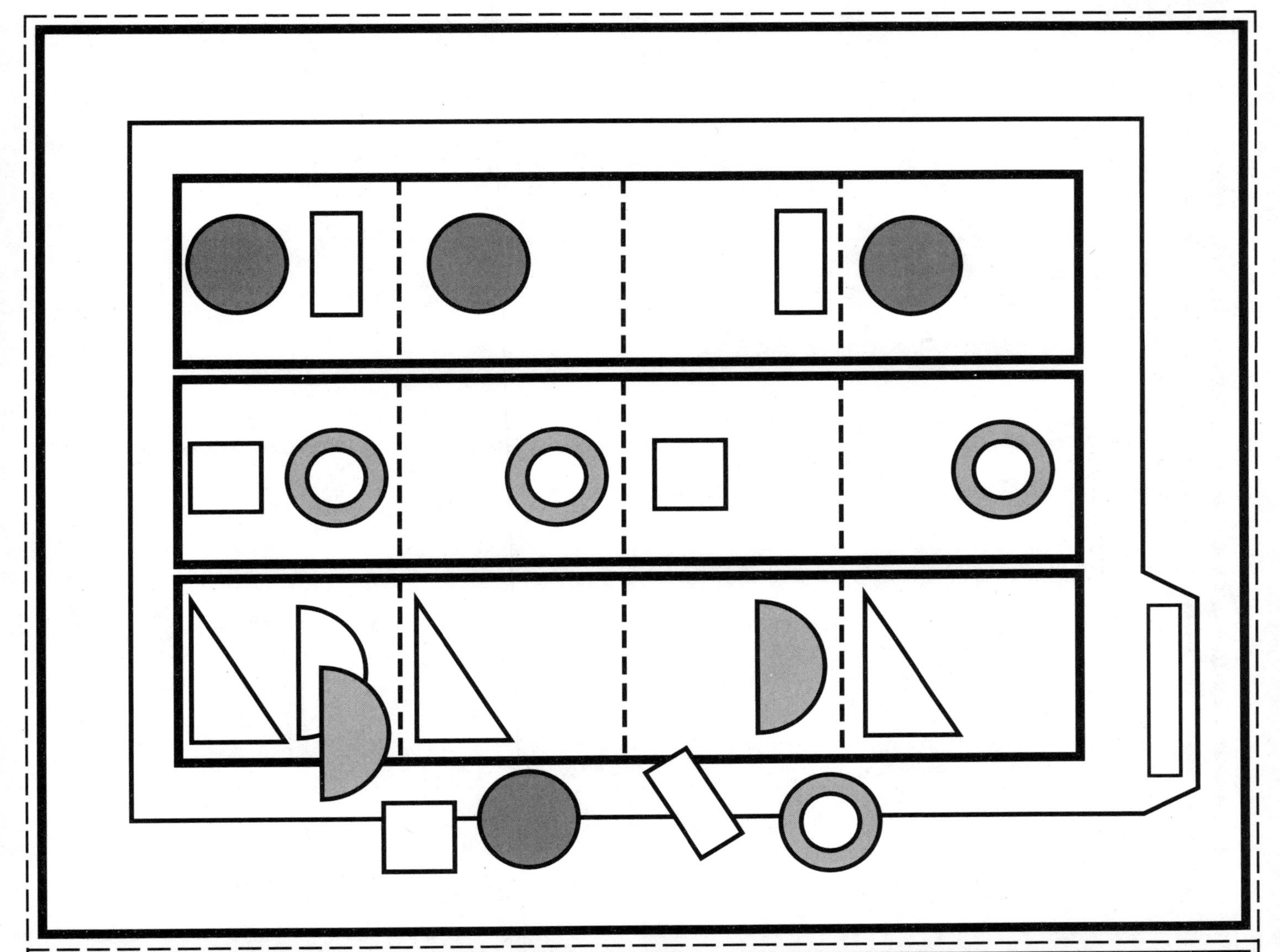

Shape Search

Shape pattern sequence

Directions: Take out the shapes and open the folder. Look at the order of the shapes in the first shape pattern in the folder. Use it to find the shapes that are missing in the row for that pattern. Find the rest of the missing shapes. Use the different shapes to make your own repeated patterns.

A book to read: Red Bear's Fun with Shapes by Bodel Rikys

Shape Search

Shape Search

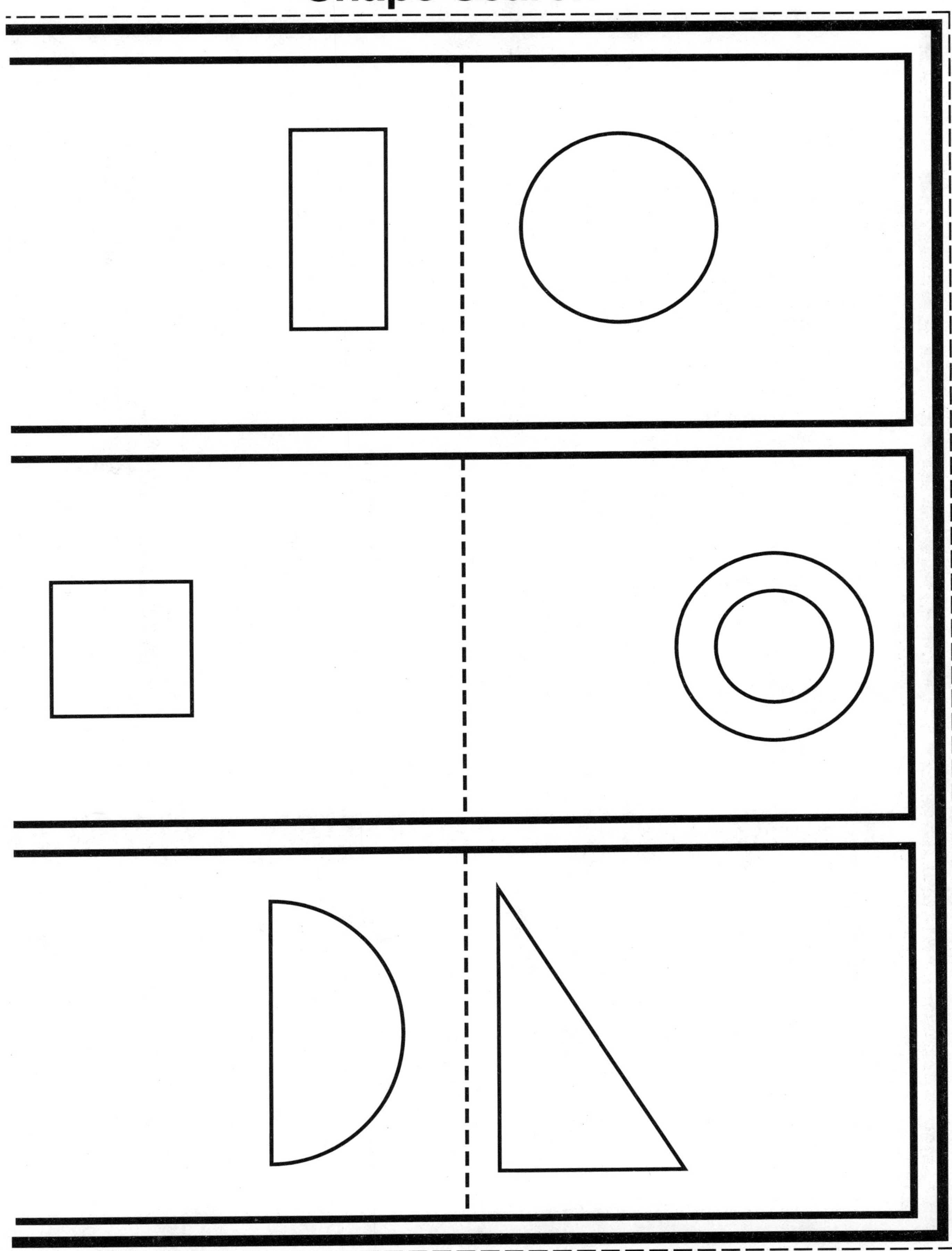

Shape Search

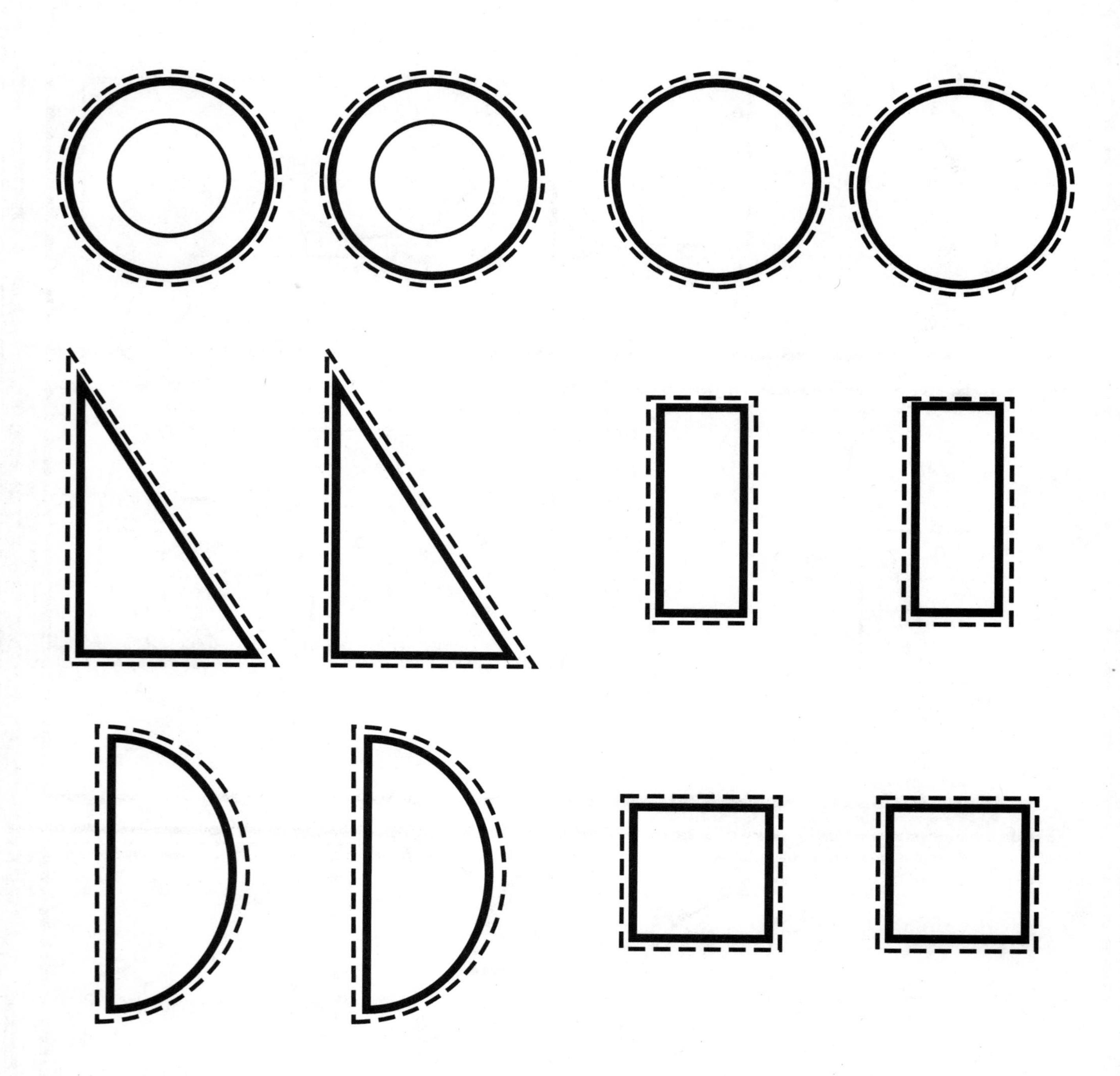

3. Pack Rat Counting

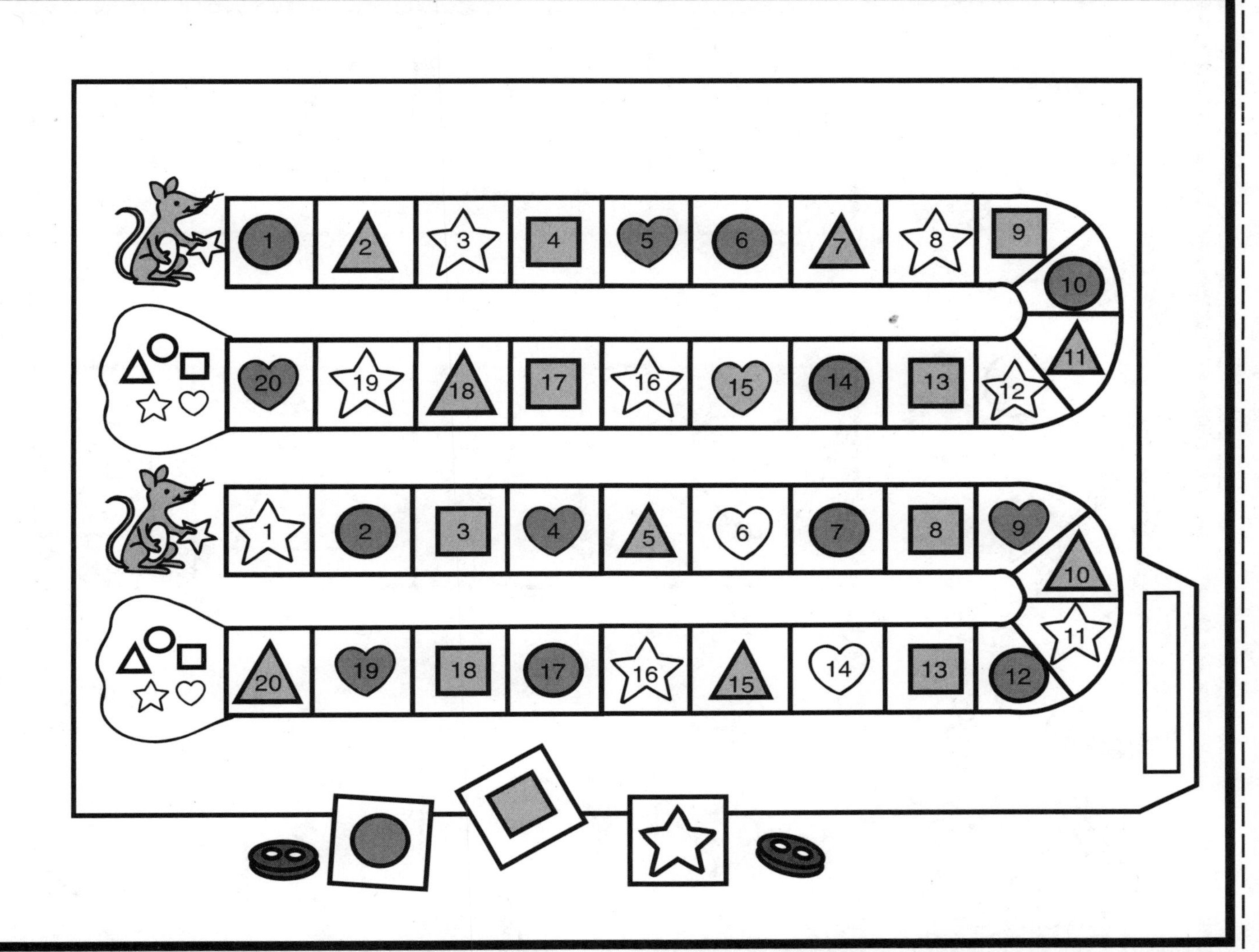

Pack Rat Counting

Counting to 20

Directions: Two children may play. Put the shape cards into groups by shape. Pick a Pack Rat trail and put your marker on the rat. Take turns throwing the die. Move the number of spaces that the die shows. Name the numbers on the spaces as you go. Take a shape card that matches the shape you stop on. Store it in Pack Rat's home at the end of the trail. Continue down the trail to 20. Count the shapes in Pack Rat's home to see who has more.

A book to read: <u>Pebble, a Pack Rat</u> by Edna Miller

Pack Rat Counting

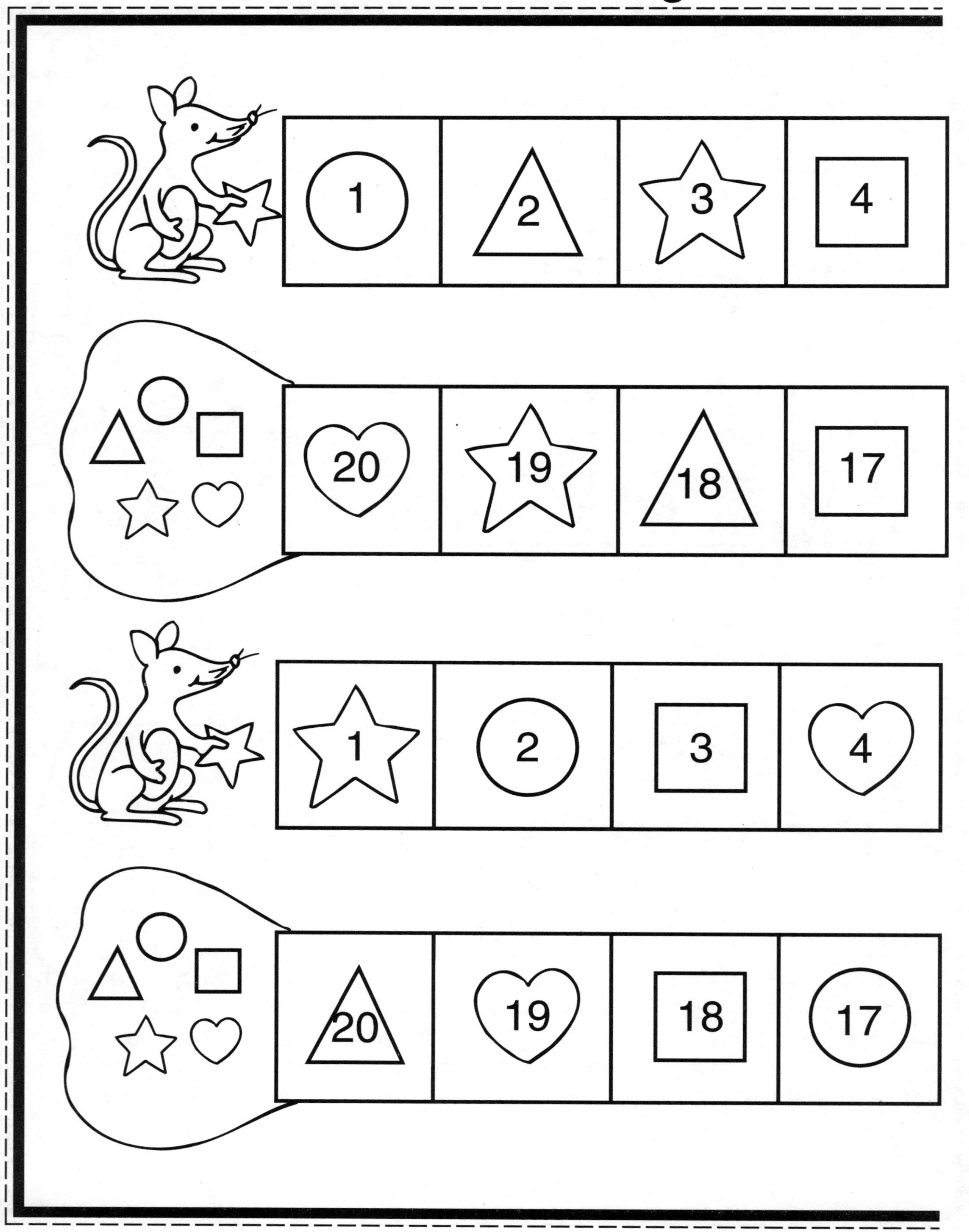

Pack Rat Counting

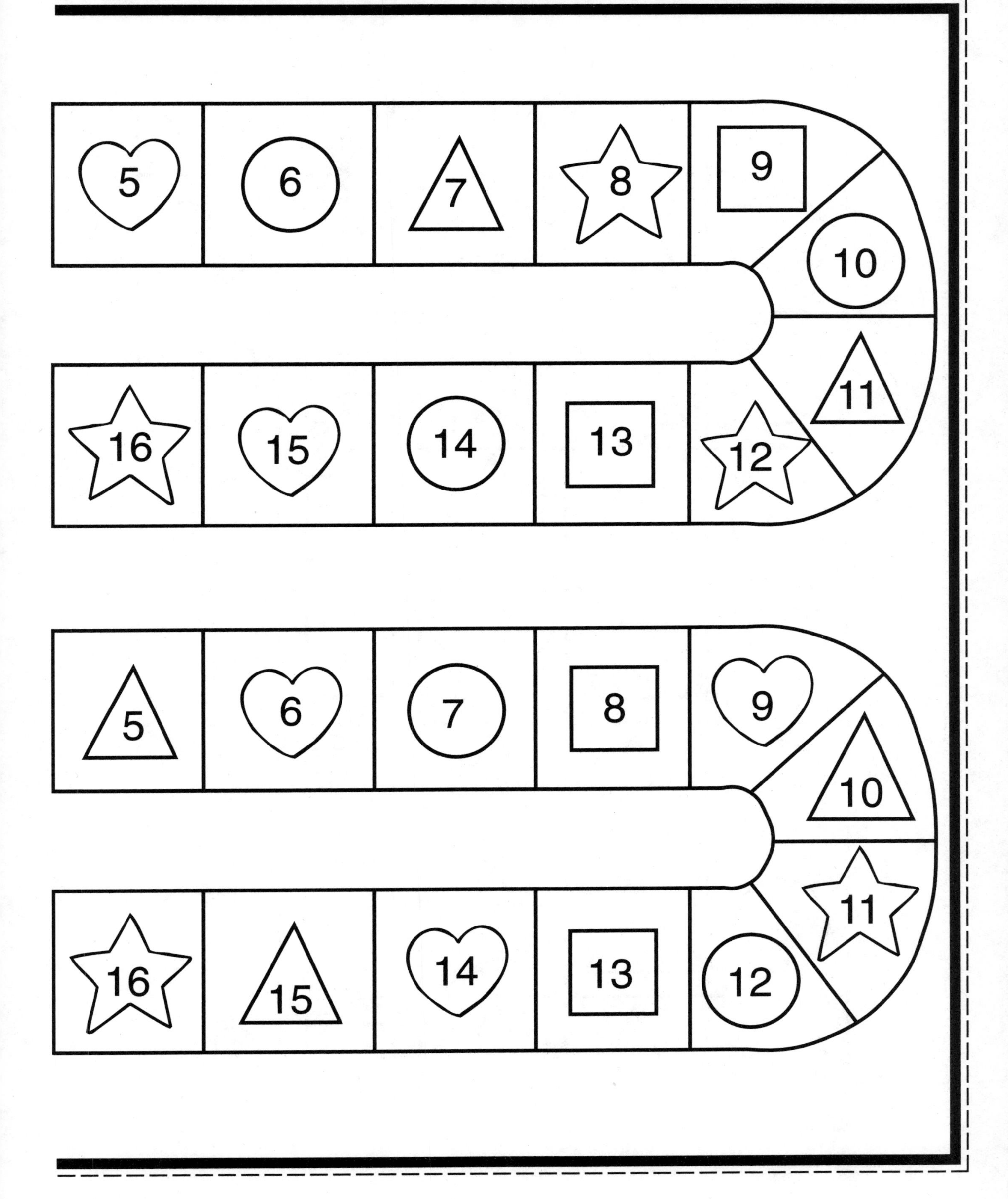

Pack Rat Counting

4. Turtle and Hare

Turtle and Hare

Counting to 50

Directions: Two children may play. Take out the markers, spinner, and trophy. Choose to be the hare or the turtle and put your marker on Start. Take turns spinning the spinner and moving your marker that number of spaces on the race track. Count each number you land on out loud. The player who gets to 50 first gets the trophy.

A book to read: Turtle and Rabbit by Valjean McLenighan

Turtle and Hare

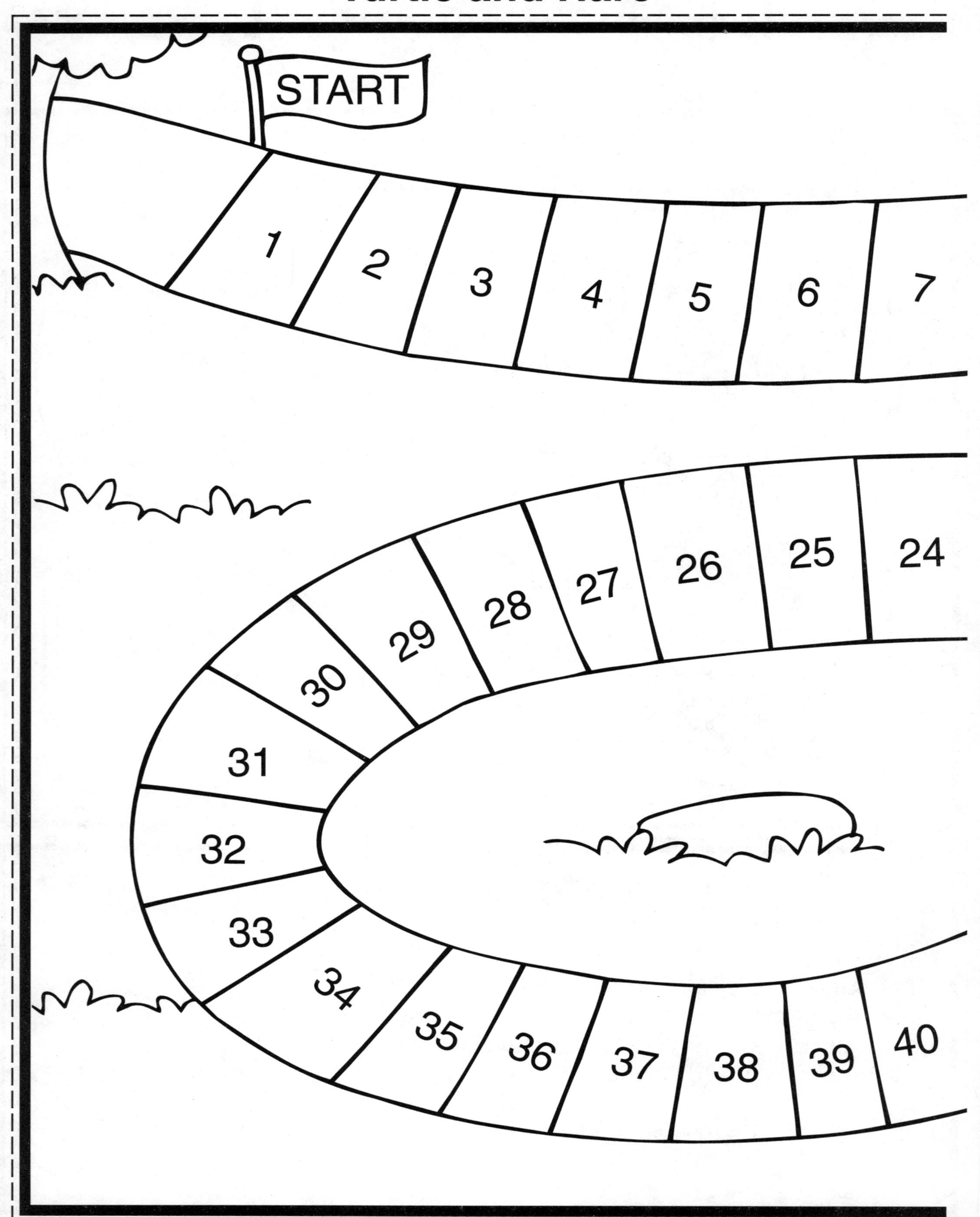

Turtle and Hare

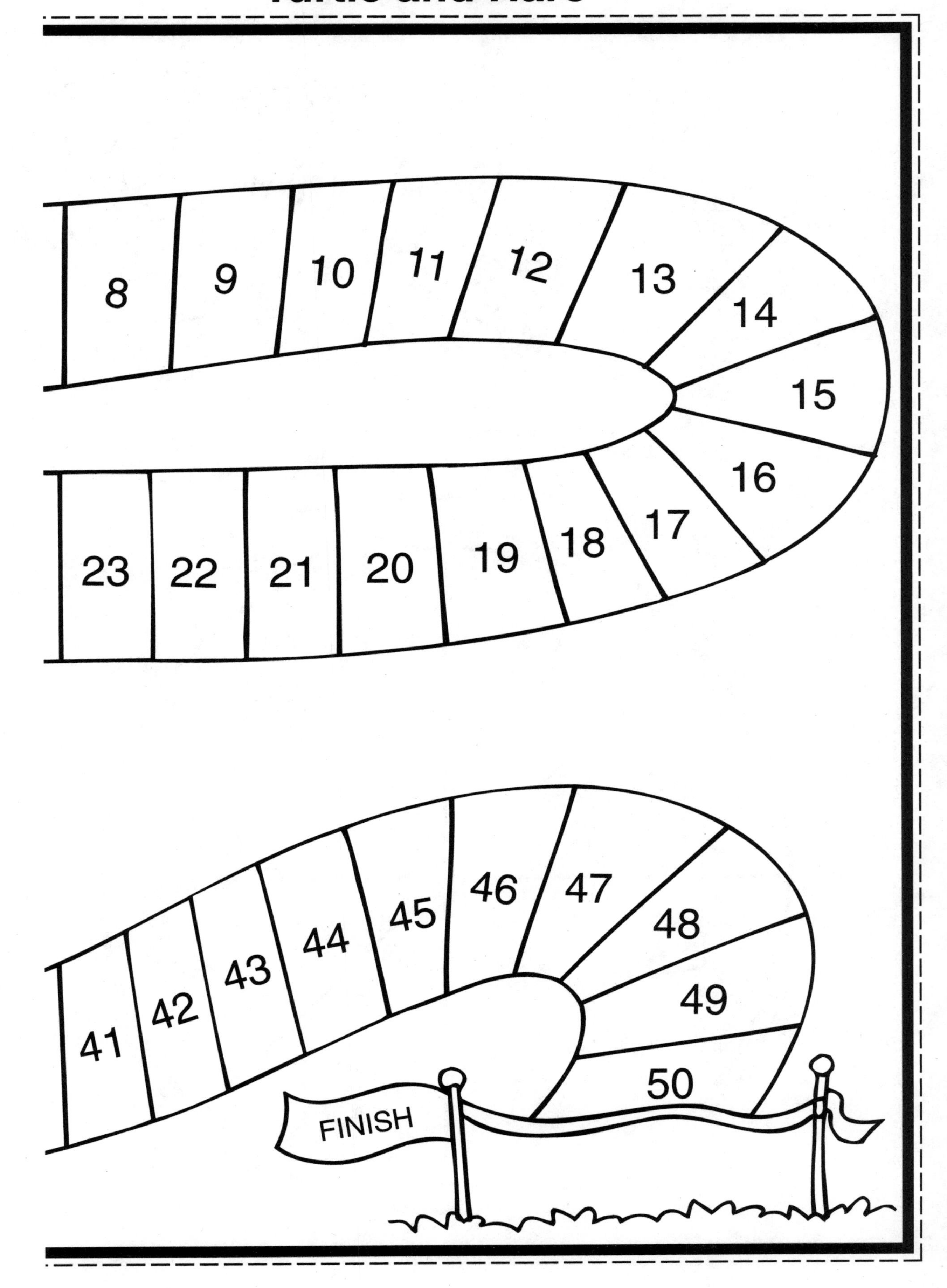

Turtle and Hare

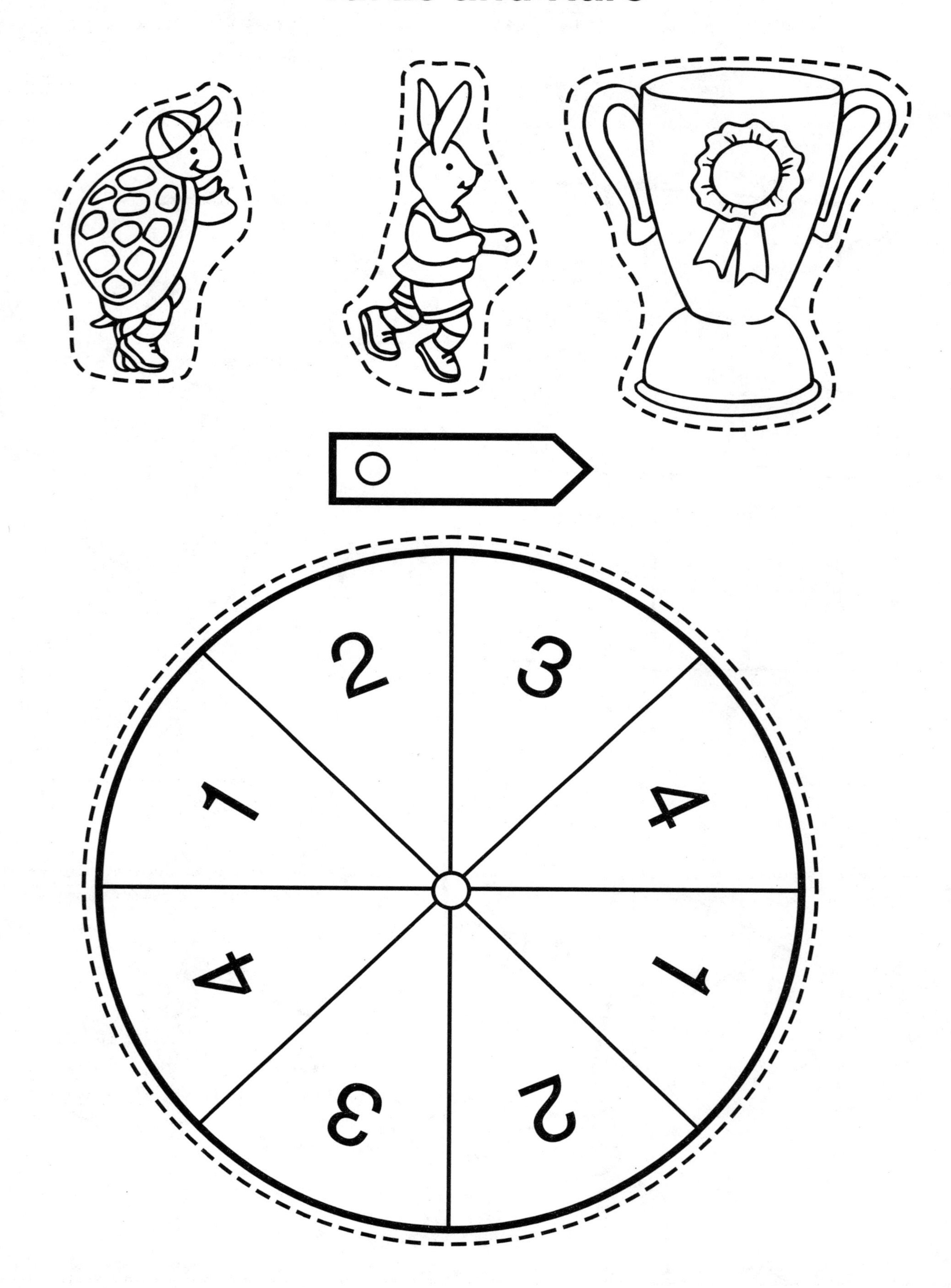

5. Tracking with Cars

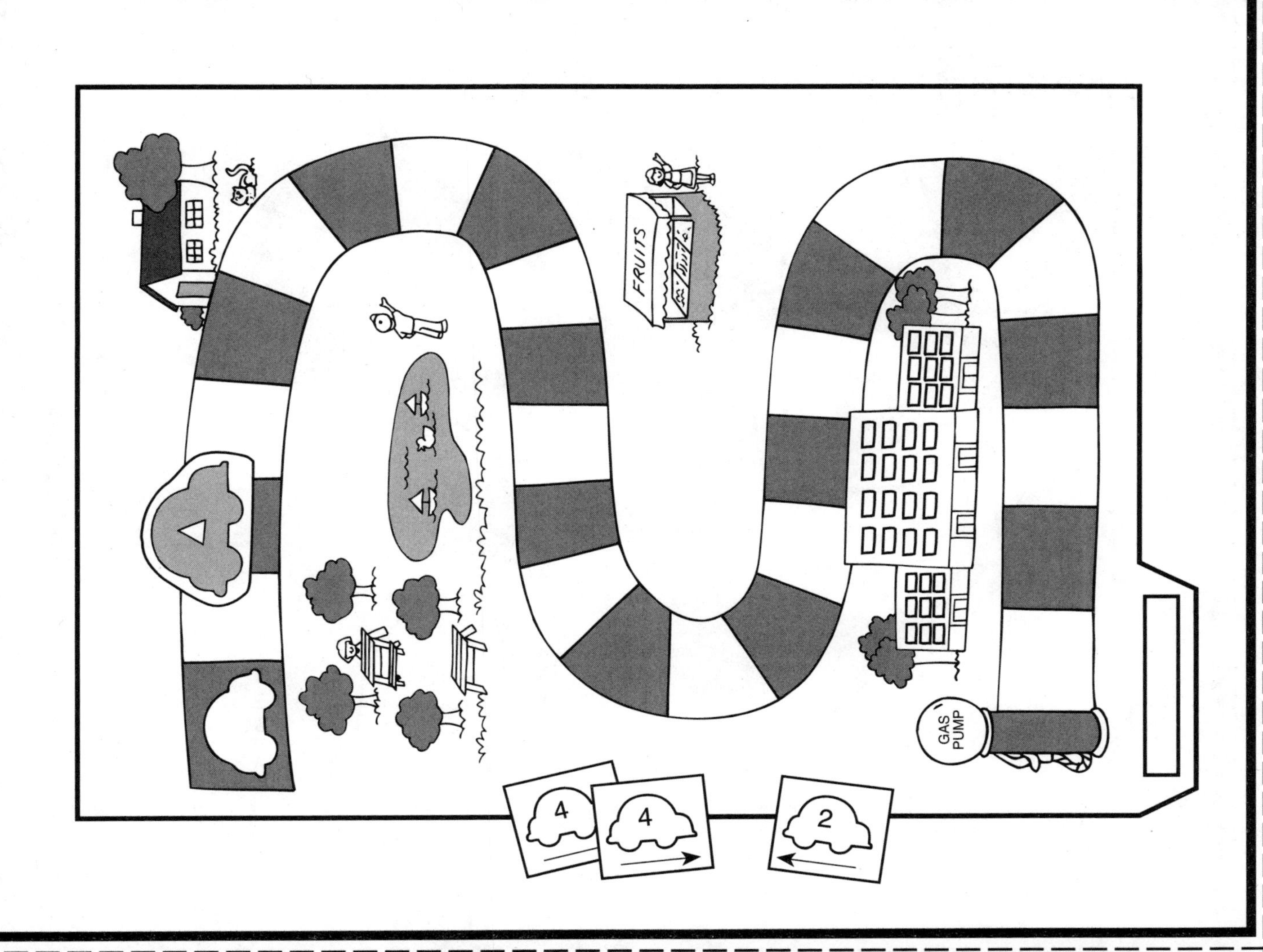

Tracking with Cars

Left-right directional skills

Directions: Three children may play. Put the cards face down. Pick a card in turn and go that number of spaces in the direction the arrow points. Put your marker on the space you stop on. Use the cards over if you need to. See who can get to the gas pump first.

A book to read: If I Drove a Car by Miriam Young

Tracking with Cars

Tracking with Cars

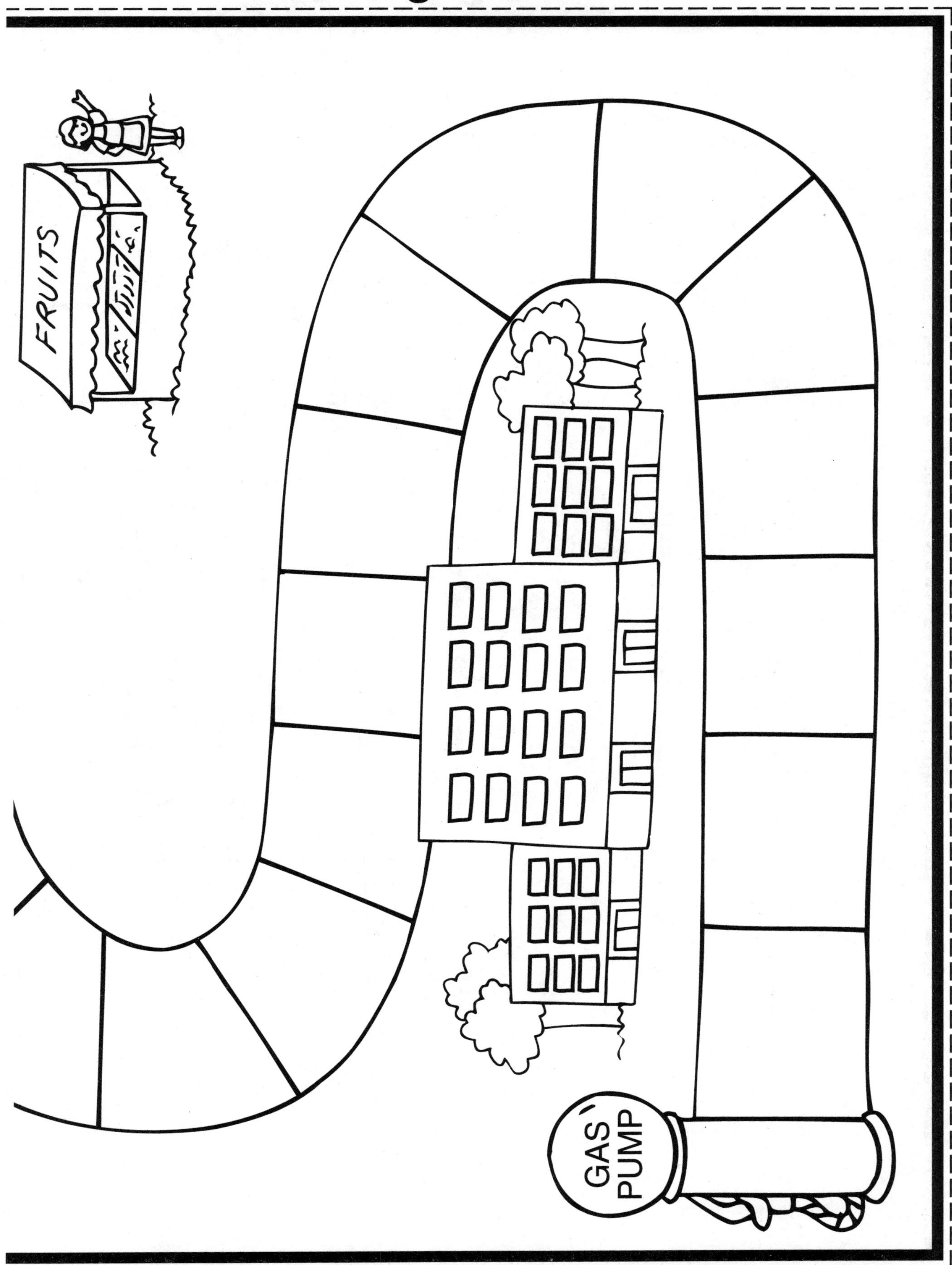

Tracking with Cars

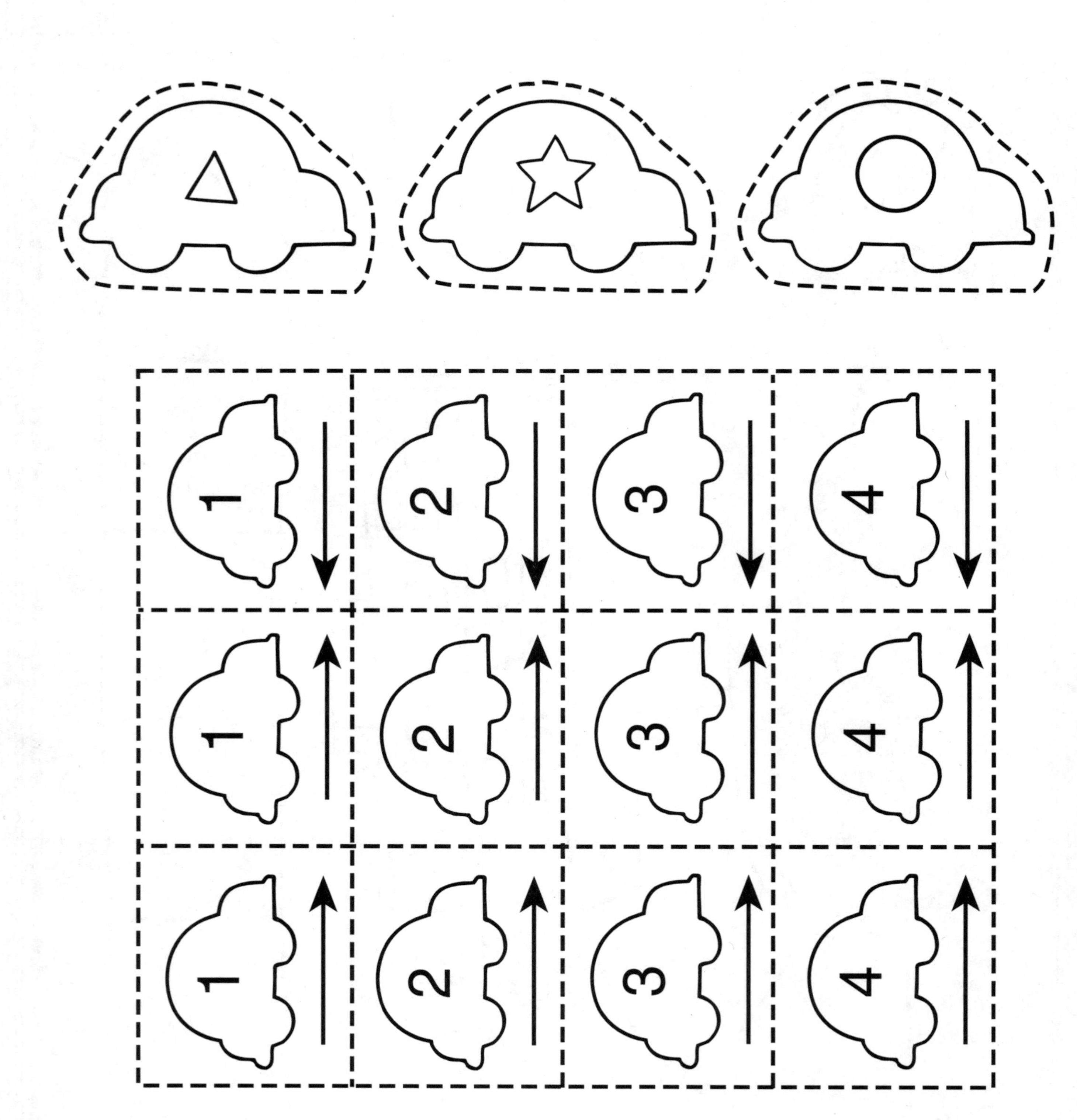

Mr. and Mrs. Kangaroo's Pockets

Classification

Directions: Two children may play. Choose either Mr. Kangaroo or Mrs. Kangaroo. Put the picture cards face down. Pick a card and name the picture. If the card goes with one of the pockets on your kangaroo, put it in the pocket. (Example: All tools go into the pocket showing a hammer.) If the card doesn't match, put it back with the other cards. Fill the pockets with the correct matches. See who can find all the cards for his or her kangaroo's pockets first.

A book to read: Katy No Pocket by Emmy Payne

Mr. and Mrs. Kangaroo's Pockets

Mr. and Mrs. Kangaroo's Pockets

Mr. and Mrs. Kangaroo's Pockets

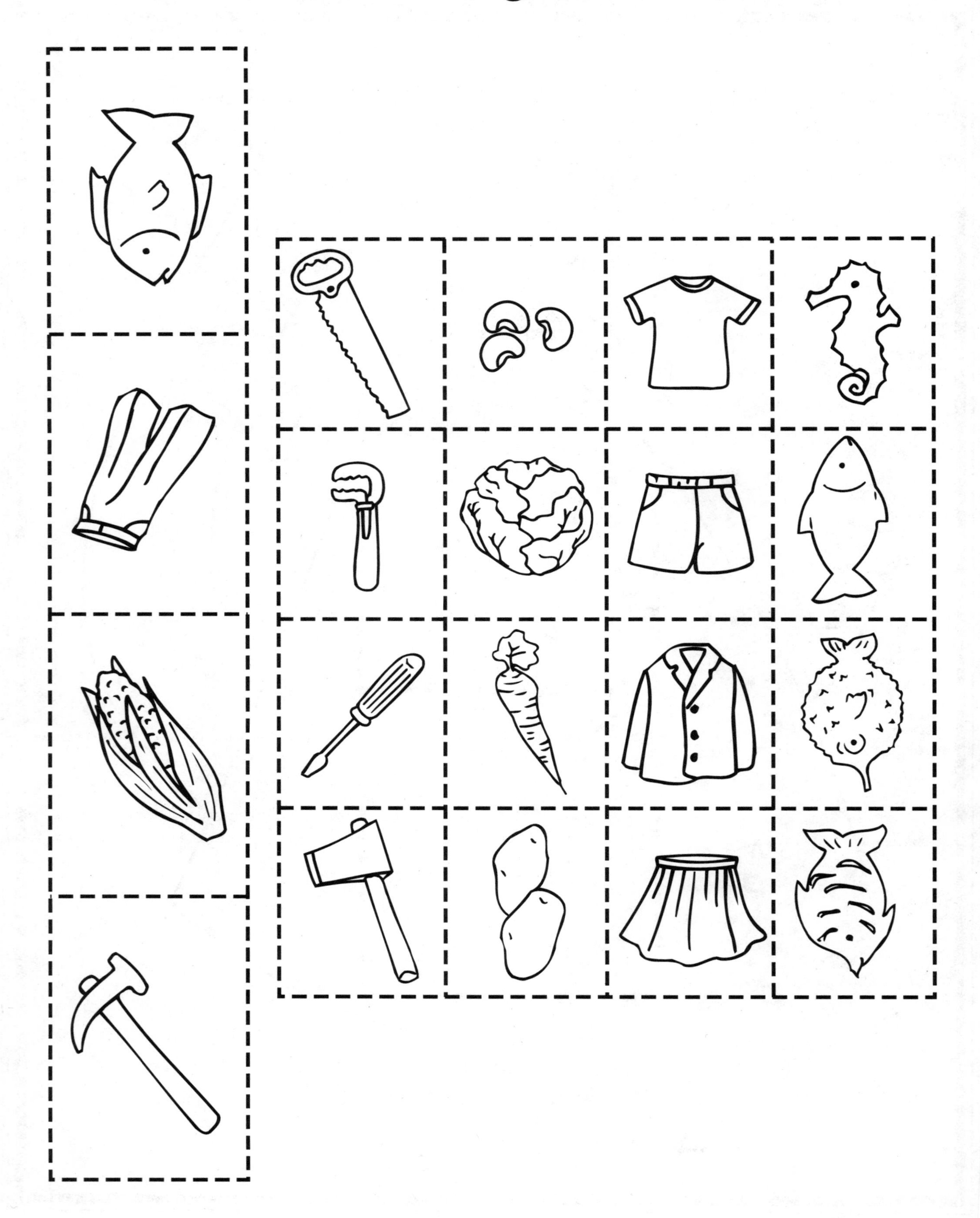

7. Big Fish, Little Fish

Big Fish, Little Fish

Size match—big/small

Directions: Two children may play. Take out the little fish cards and open the folder. Put the little fish cards face down. Choose which side of the folder you want to play on. Pick a fish card in turn and see if it matches one of your big fish. If it does, put the card on the big fish. If it doesn't, put the card back with the other cards. See who can find all the big fish/little fish matches first.

A book to read: Fish Is Fish by Leo Lionni

Big Fish, Little Fish

Big Fish, Little Fish

Big Fish, Little Fish

8. Boxes and Balls

Boxes and Balls

Size sequence—big/small

Directions: Two children may play. Take out the box and ball shapes and place them on the playing area. Choose either the row of boxes or the row of balls on the folder to play. Take turns picking up the shapes. Put them on the matching boxes or balls from biggest to smallest. See who can make all the big-to-small matches first.

A book to read: Boxes, Boxes! by Leonard Everett Fisher

Boxes and Balls

9. Snakes in the Grass

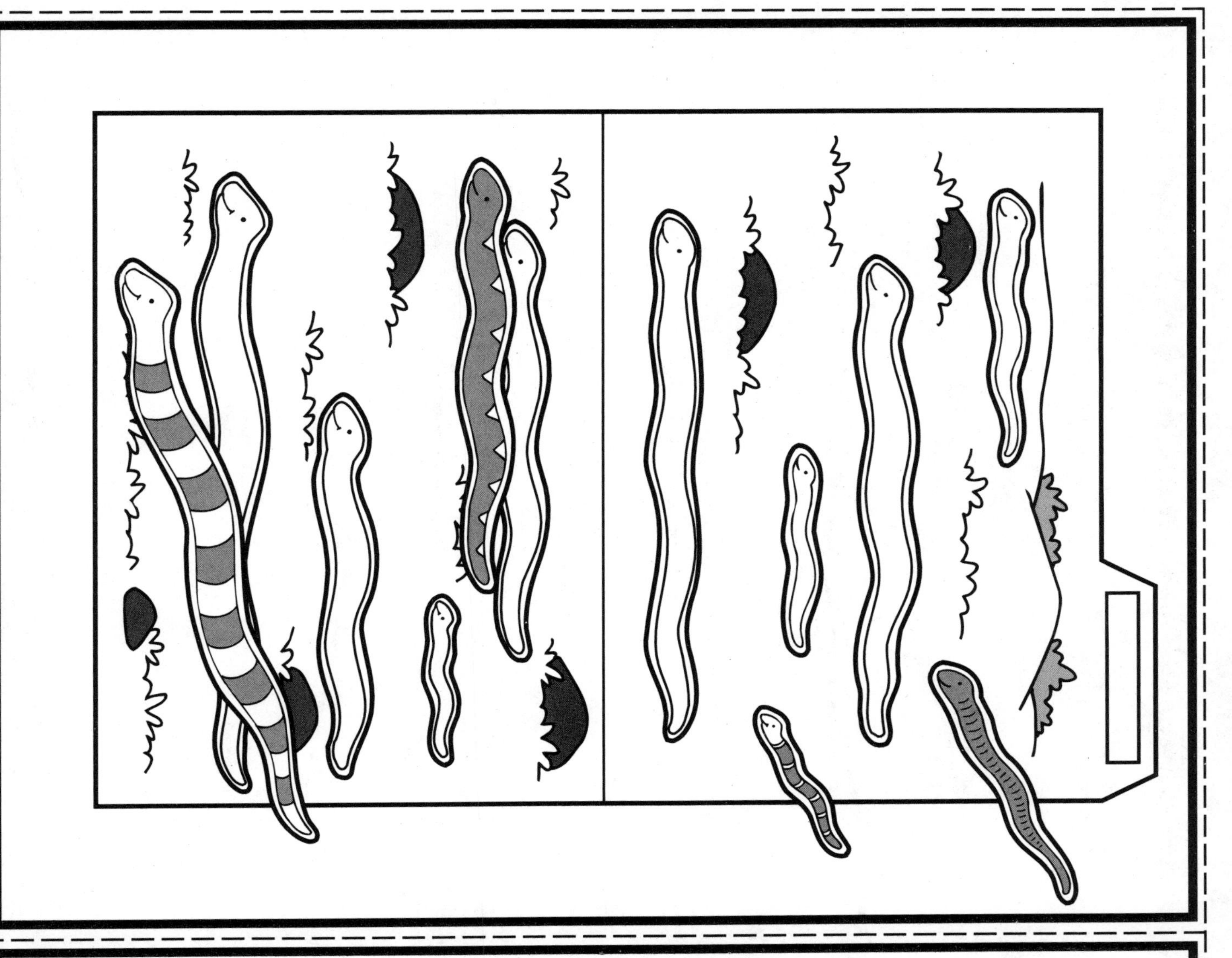

Snakes in the Grass

Length—short/long

Directions: Two children may play. Take out the snakes and open the folder. Take turns finding the snakes that match the long and short snakes on the folder. Put the matches on the snakes. Next, have fun putting the snakes in order, from the longest to the shortest or the shortest to the longest.

A book to read: The Singing Snake by Stefan Czernecki and Timothy Rhodes

Snakes in the Grass

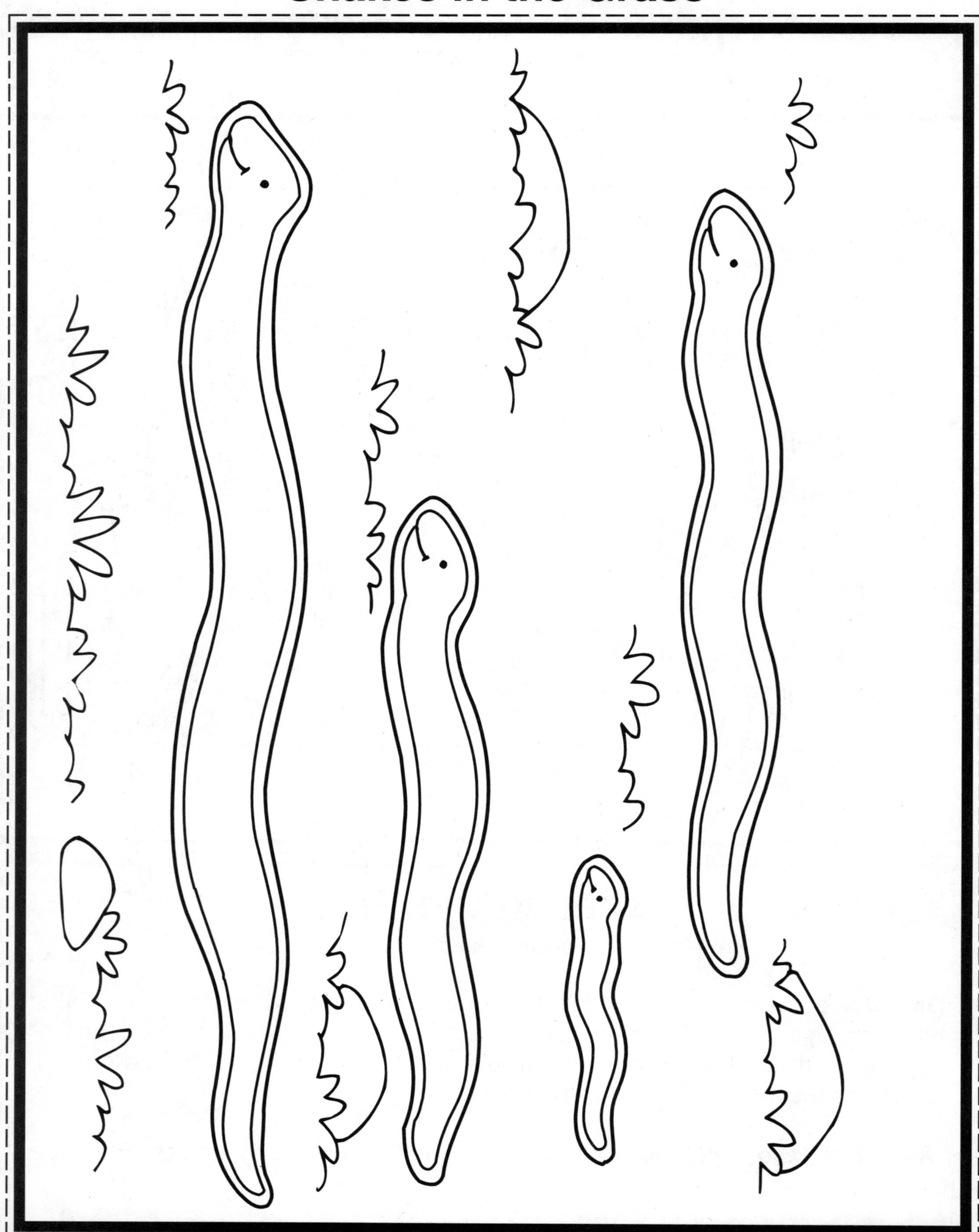

Snakes in the Grass

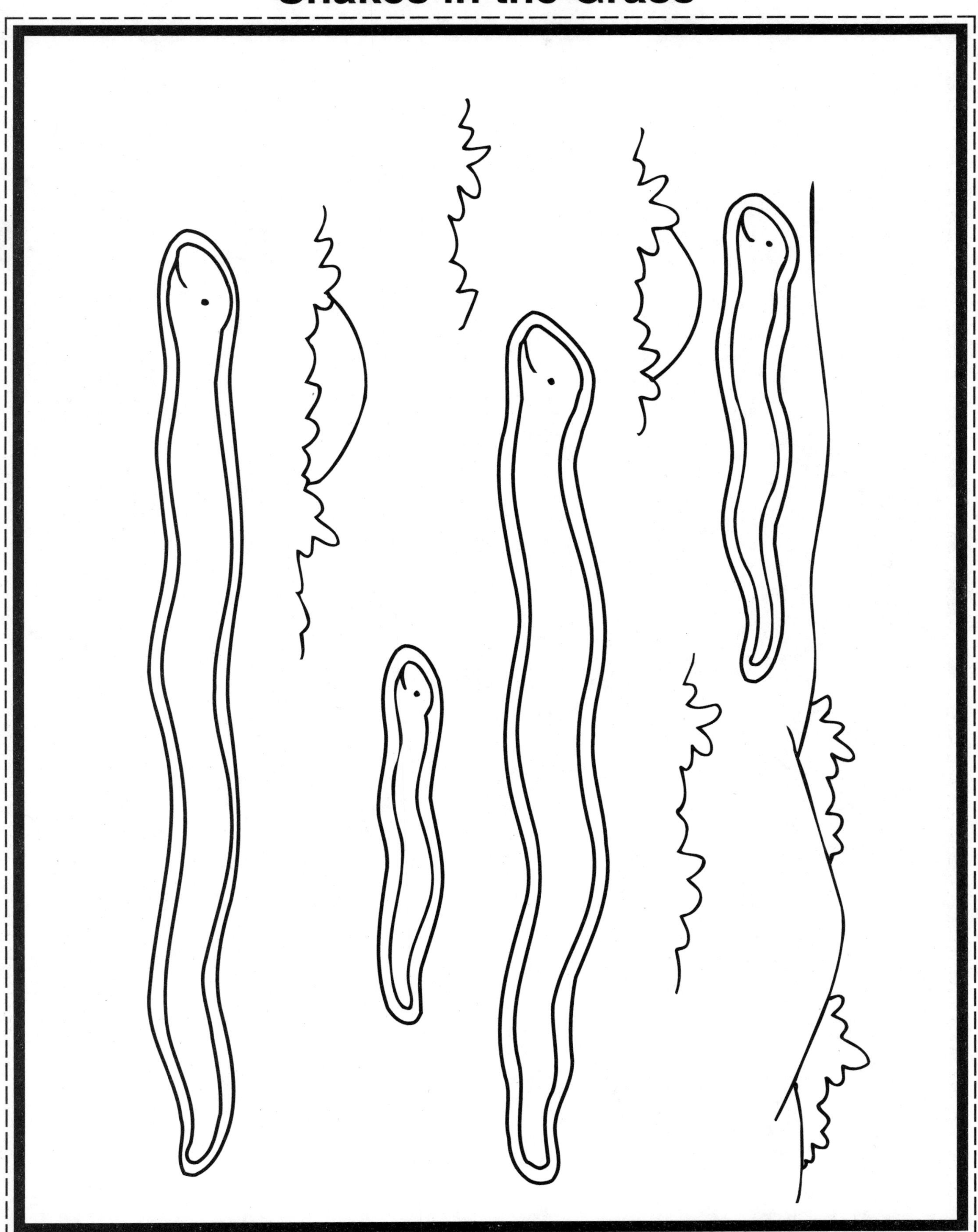

Snakes in the Grass

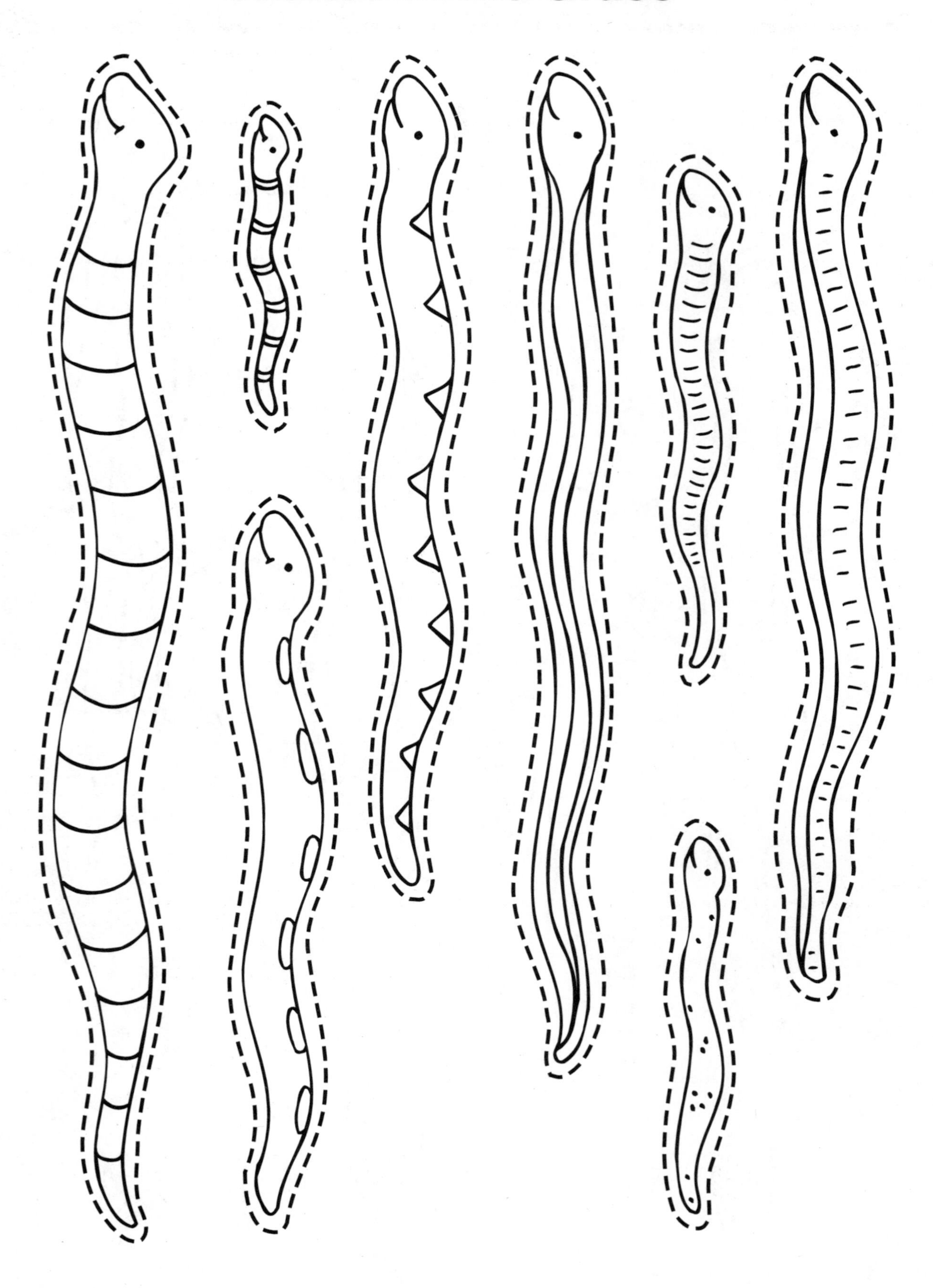

10. Duck and Bear Look-alikes

Duck and Bear Look-alikes

Concept of sameness

Directions: Two children may play. Take out the bear and duck cards and open the folder. Put the cards face down. Choose to play either the ducks or the bears. Pick a card in turn. See if it looks like one of your bears or ducks. Put the look-alike on the row with the matching bear or duck. If the card is not a match, put it back with the other cards. See who can find all of his or her look-alikes first.

A book to read: Just Like Me by Miriam Schlen

Duck and Bear Look-alikes

Duck and Bear Look-alikes

Duck and Bear Look-alikes

11. Bones

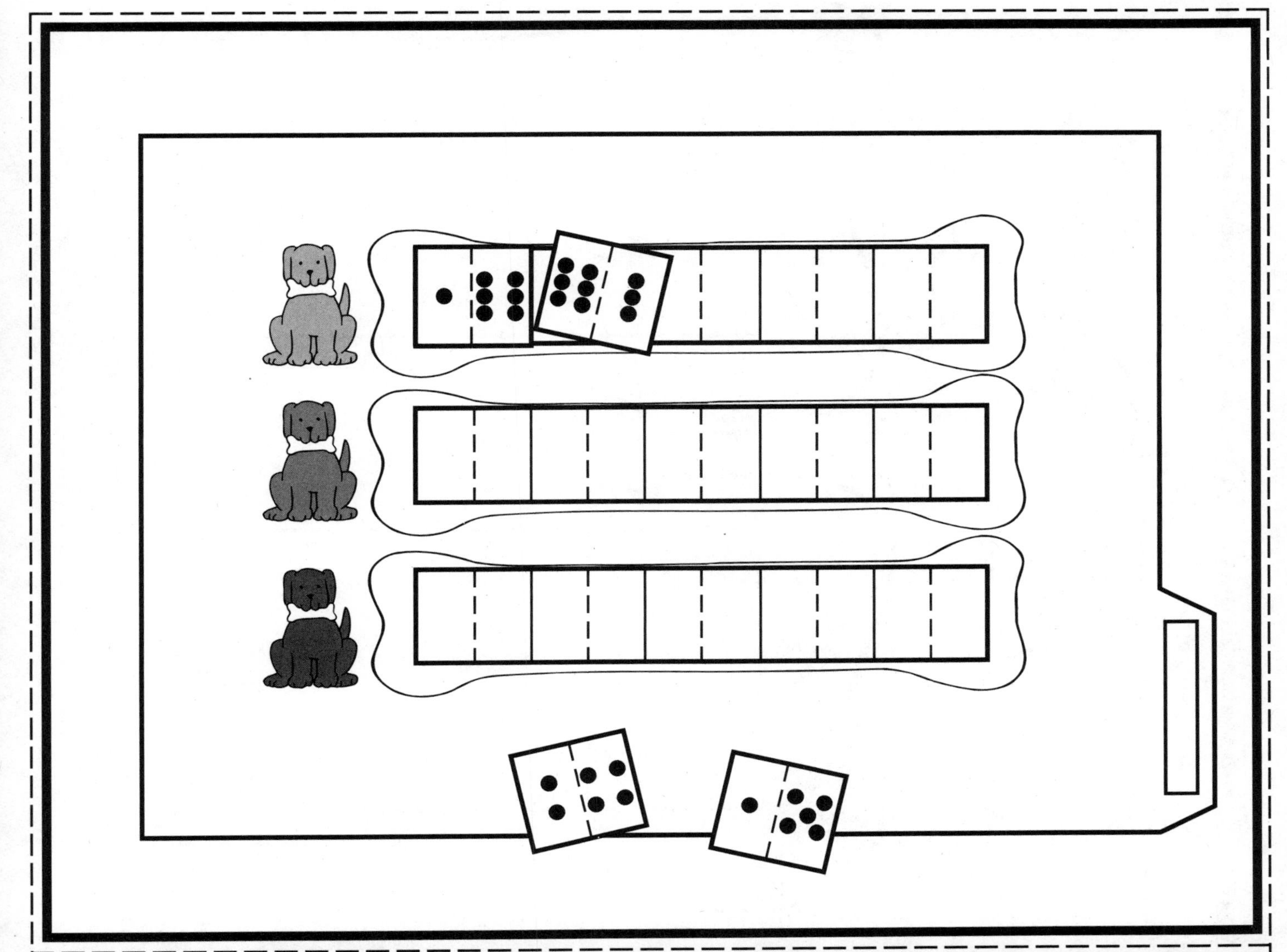

Bones

Number pattern sequence

Directions: Three children may play. Put the domino cards face down. Pick a bone to play on. Pick a domino card in turn and place it on your bone to start. On your next turn, if the card you pick matches the last number pattern on your domino, put it on your bone. If it does not match, put the card back. Keep on matching the last number pattern on each domino. See who can find all the matching dominos on his or her bone first.

A book to read: Teeny Tiny Woman by Paul Galdone

Bones

Bones

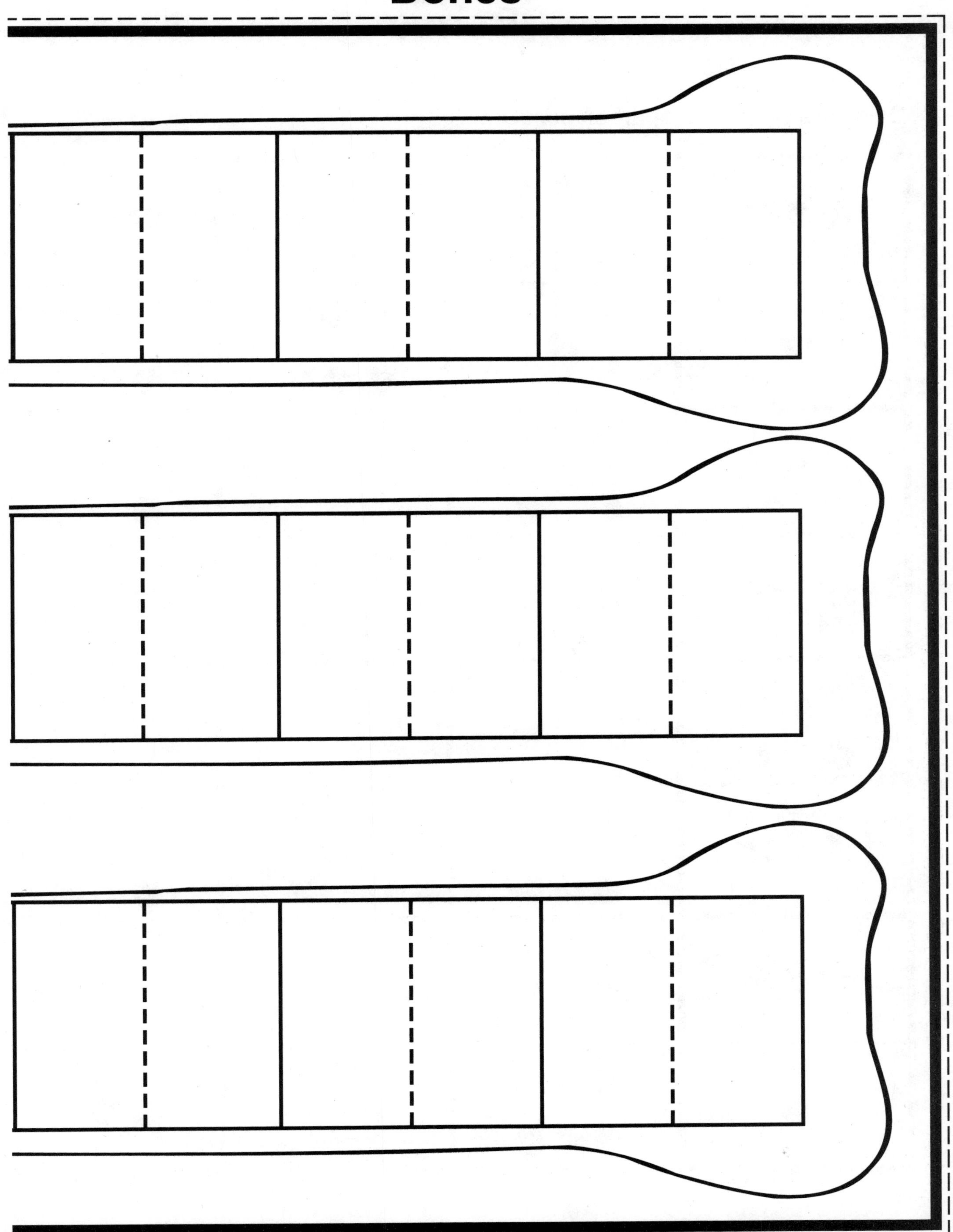

Bones

Duplicate twice. Cut apart on solid lines.

12. Mittens

Mittens

Counting/1-1 matching

Directions: Put the square counters in rows on the playing area. Choose one of the mittens to play with. Count aloud the number of objects on your mitten. It should match the number written on the mitten. Pick up the same number of counters and put one on each object on your mitten. Count aloud as you do this. Keep on matching objects on the mittens with the counters. Say how many things are in each mitten each time. Special challenge: What is the total number of objects on all the mittens? Count!

A book to read: The Mitten by Alvin Tresselt

Mittens

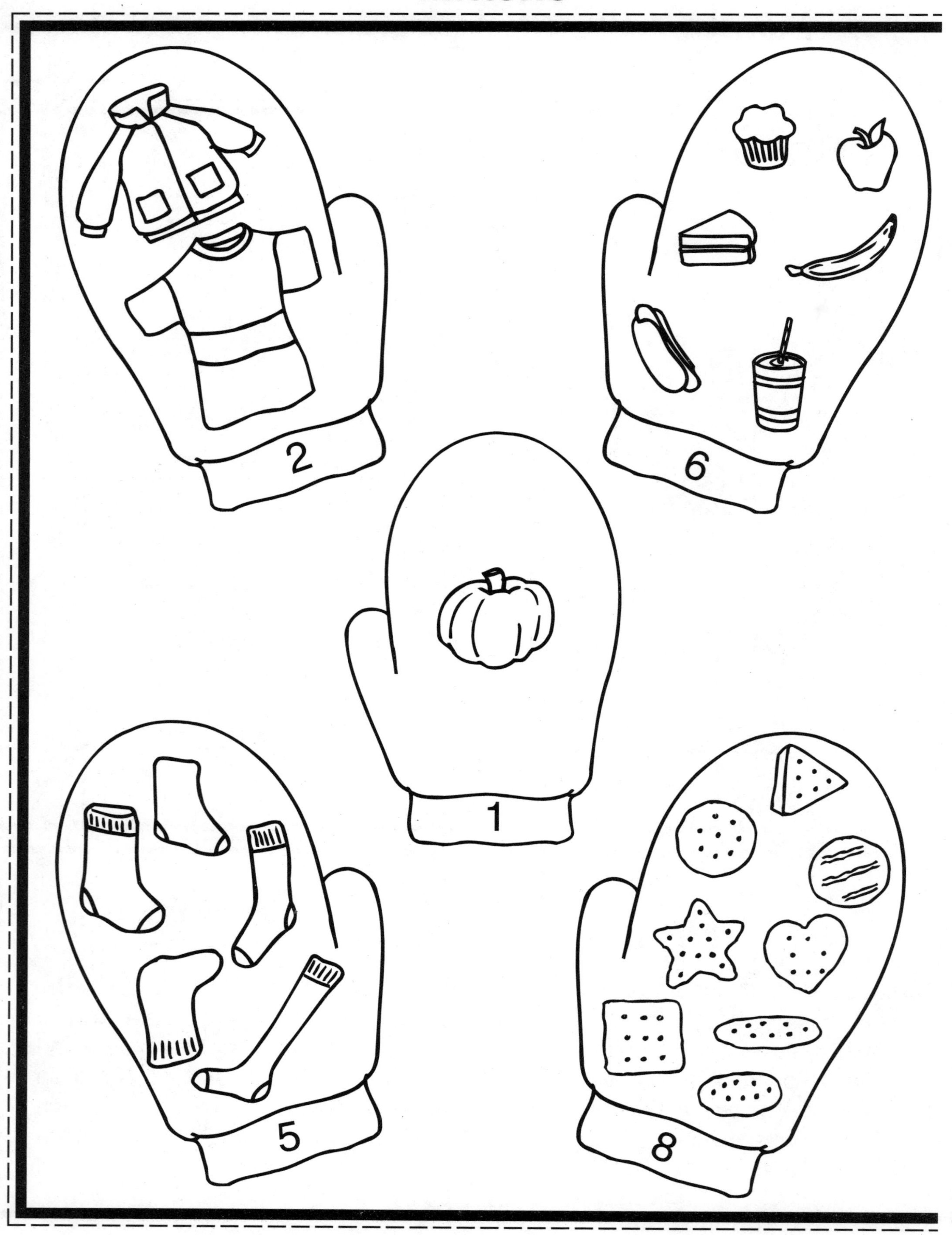

Mittens

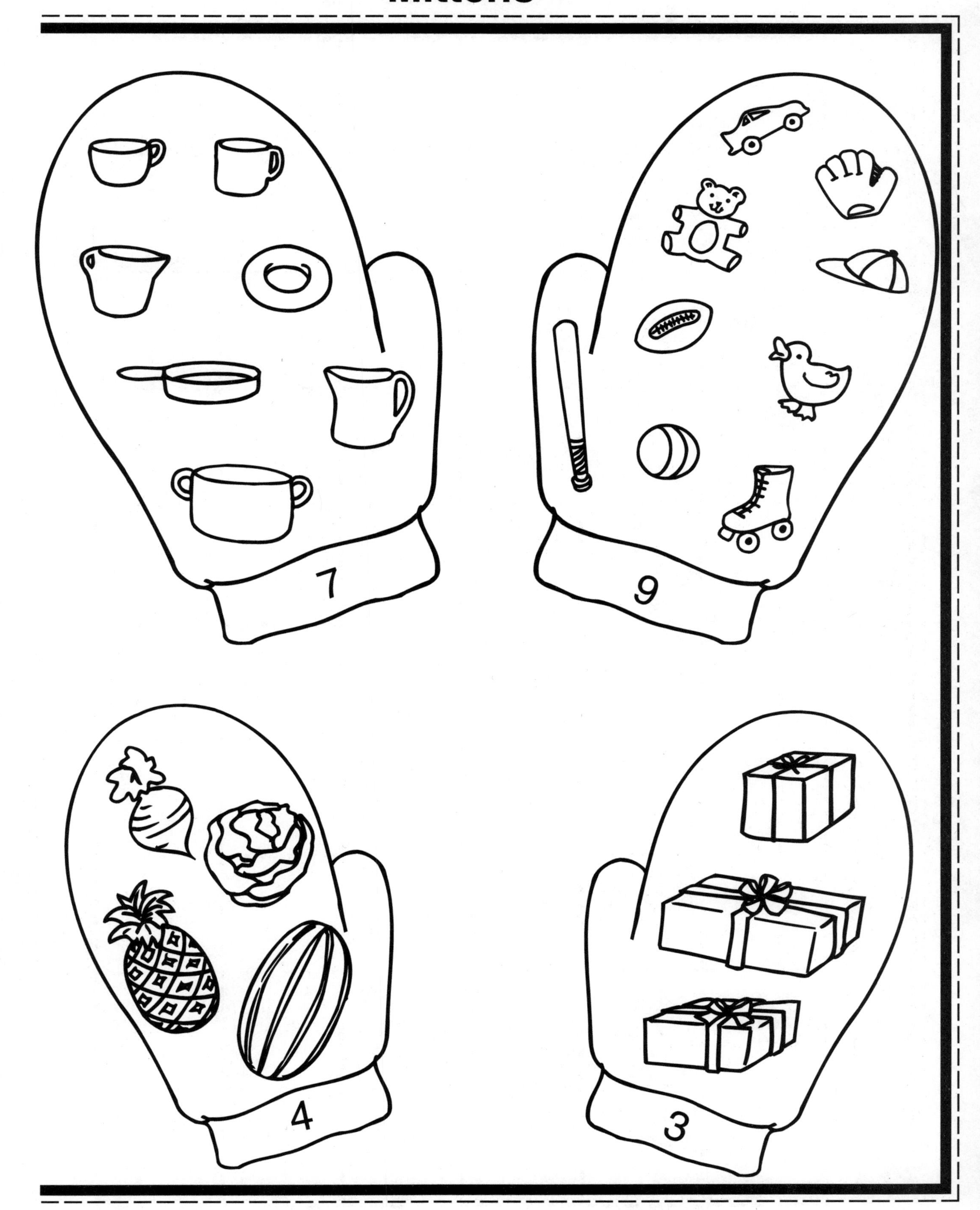

Mittens

13. Very Beary T-shirts

Very Beary T-shirts

Identifying number symbols and words

Directions: Two children may play. Take out the T-shirts and open the folder. Put the T-shirts face down in rows in the playing area. Take a T-shirt in turn. Count the dots on the shirt and read the number word. Find the bear with the same number to match the shirt. Count the number of fingers the bear is holding up to check. Put the shirt on the bear.

A book to read: We're Going on a Bear Hunt by Michael Rosen

Very Beary T-shirts

Very Beary T-shirts

Very Beary T-shirts

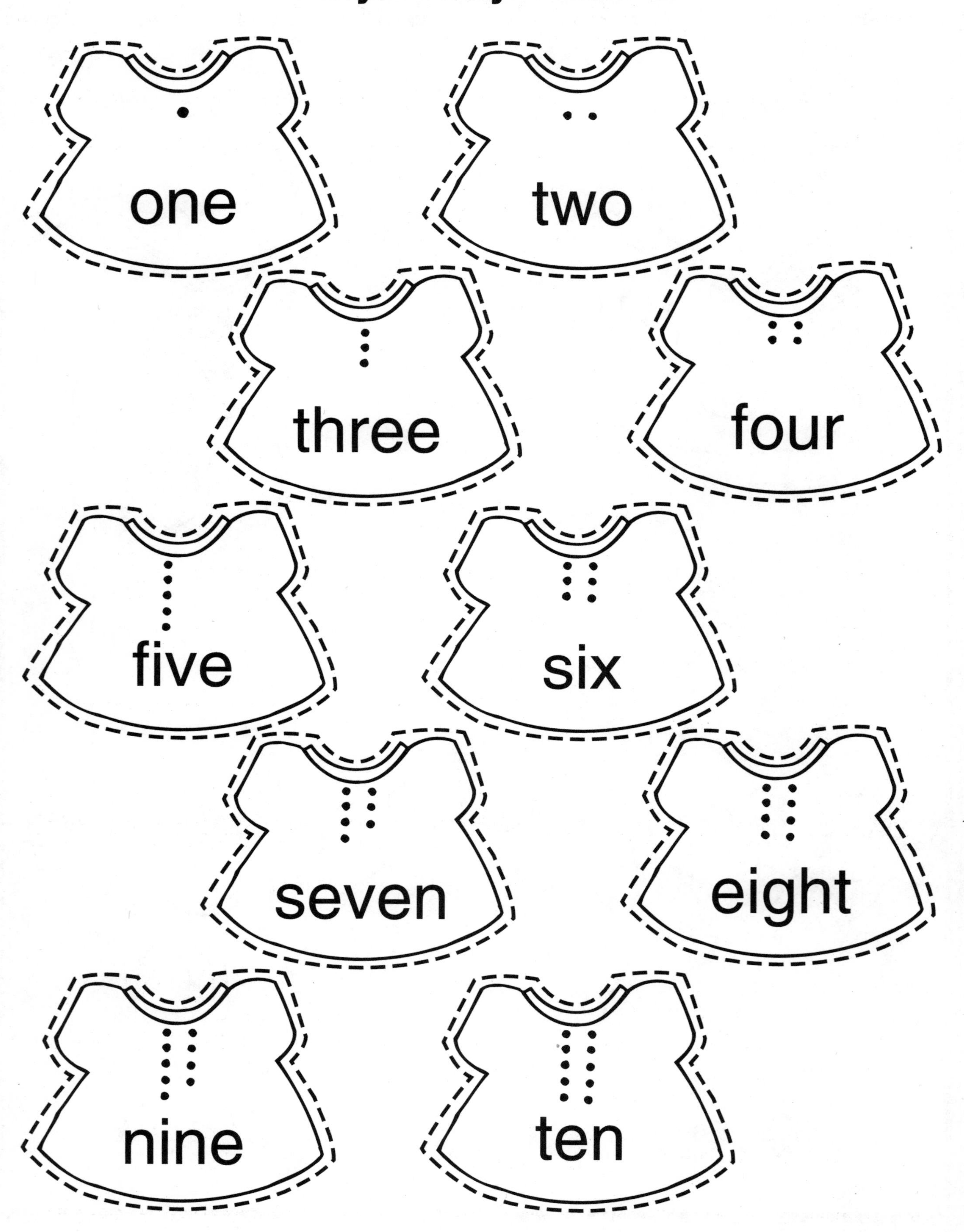

14. Mother Wolf's Quilts

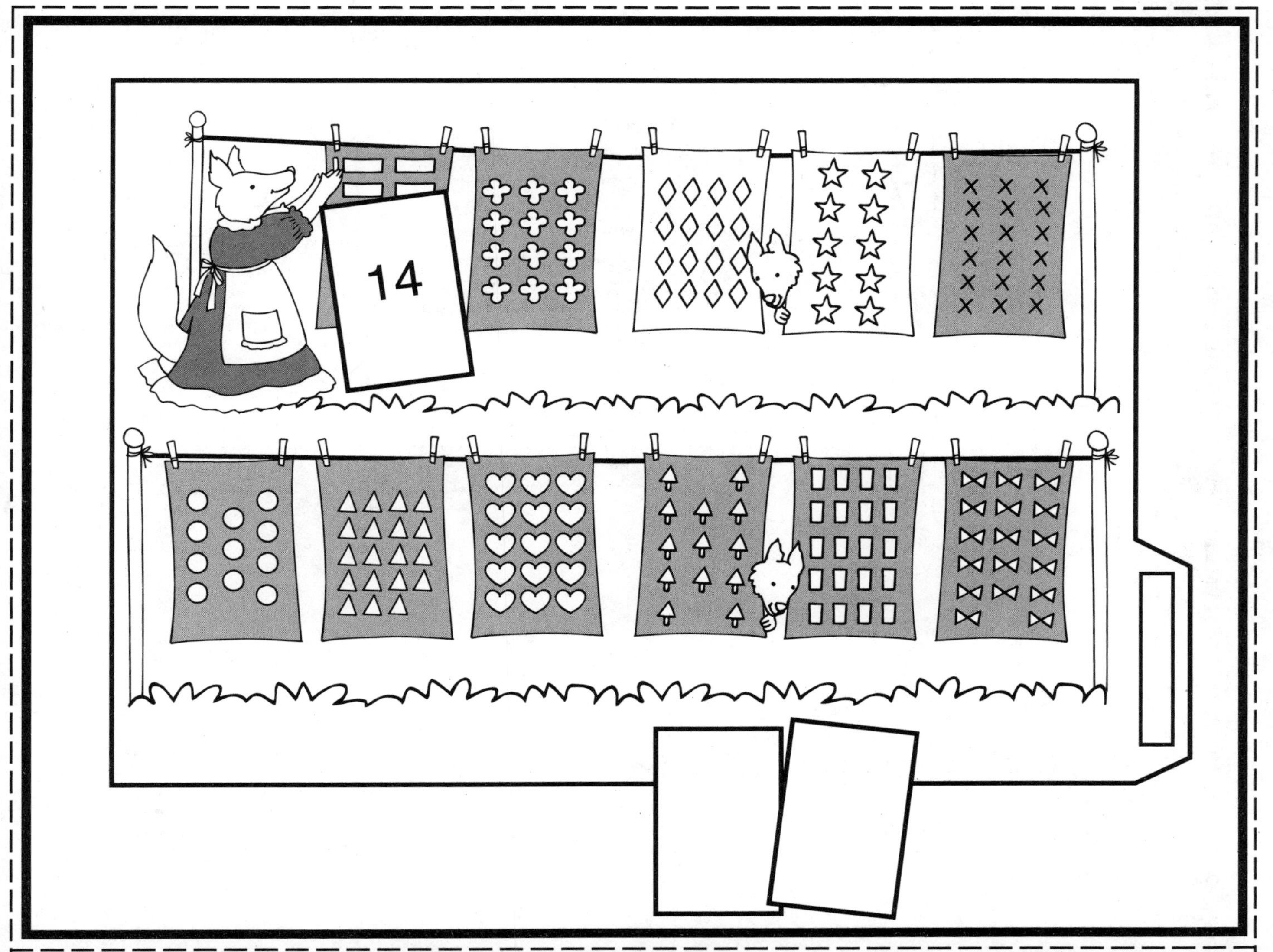

Mother Wolf's Quilts

Counting and number sets

Directions: Take out the number cards and place them number side down. Open the folder. Look at the number of shapes on each quilt. Take a number card. Find the quilt that matches the number. Place the number card on that quilt. Say the numbers as you count to check your work.

A book to read: The Patchwork Quilt by Valerie Flournoy

Mother Wolf's Quilts

Mother Wolf's Quilts

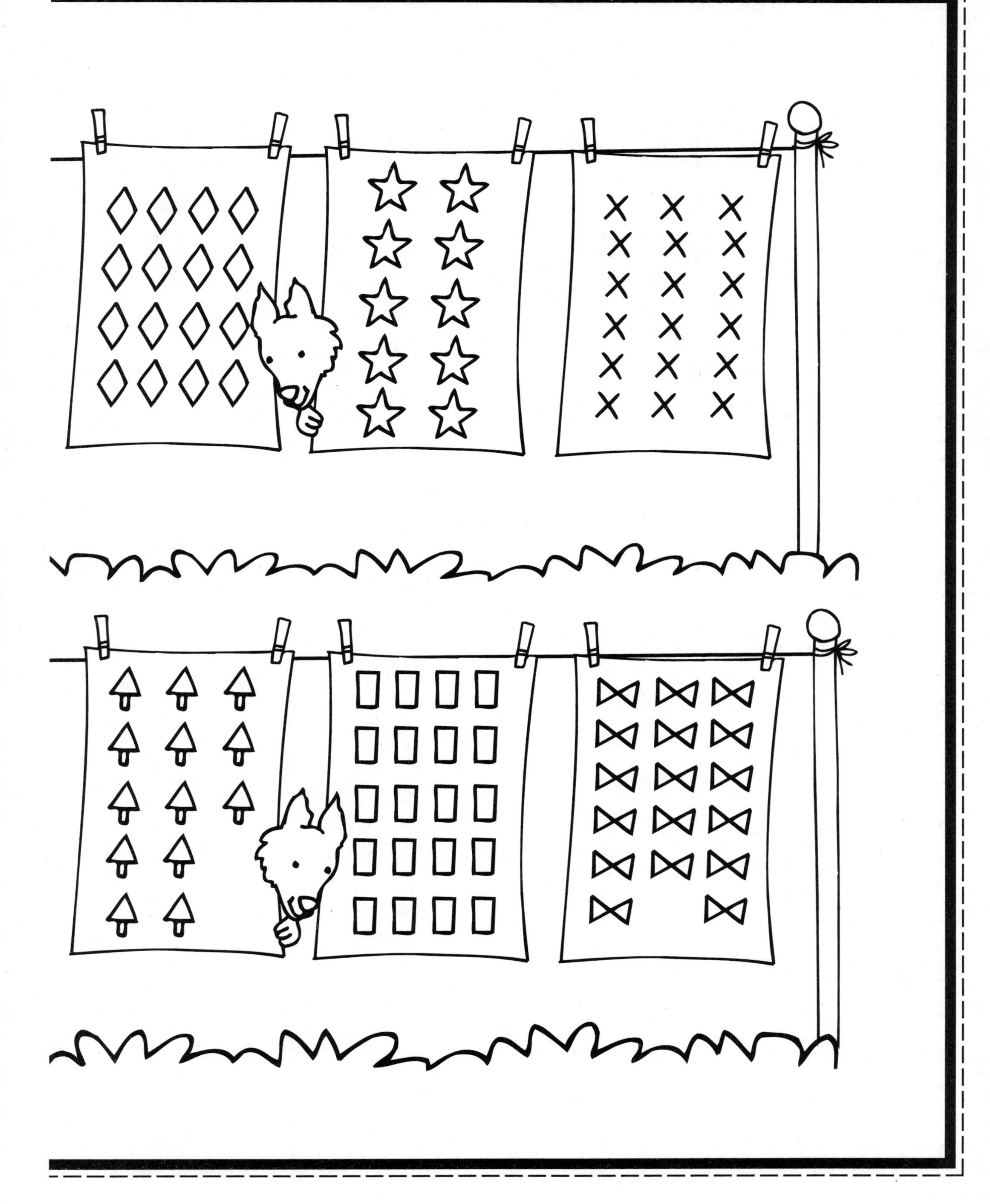

Mother Wolf's Quilts

10	11	12	13
14	15	16	17
18	19	20	

15. Space Walk

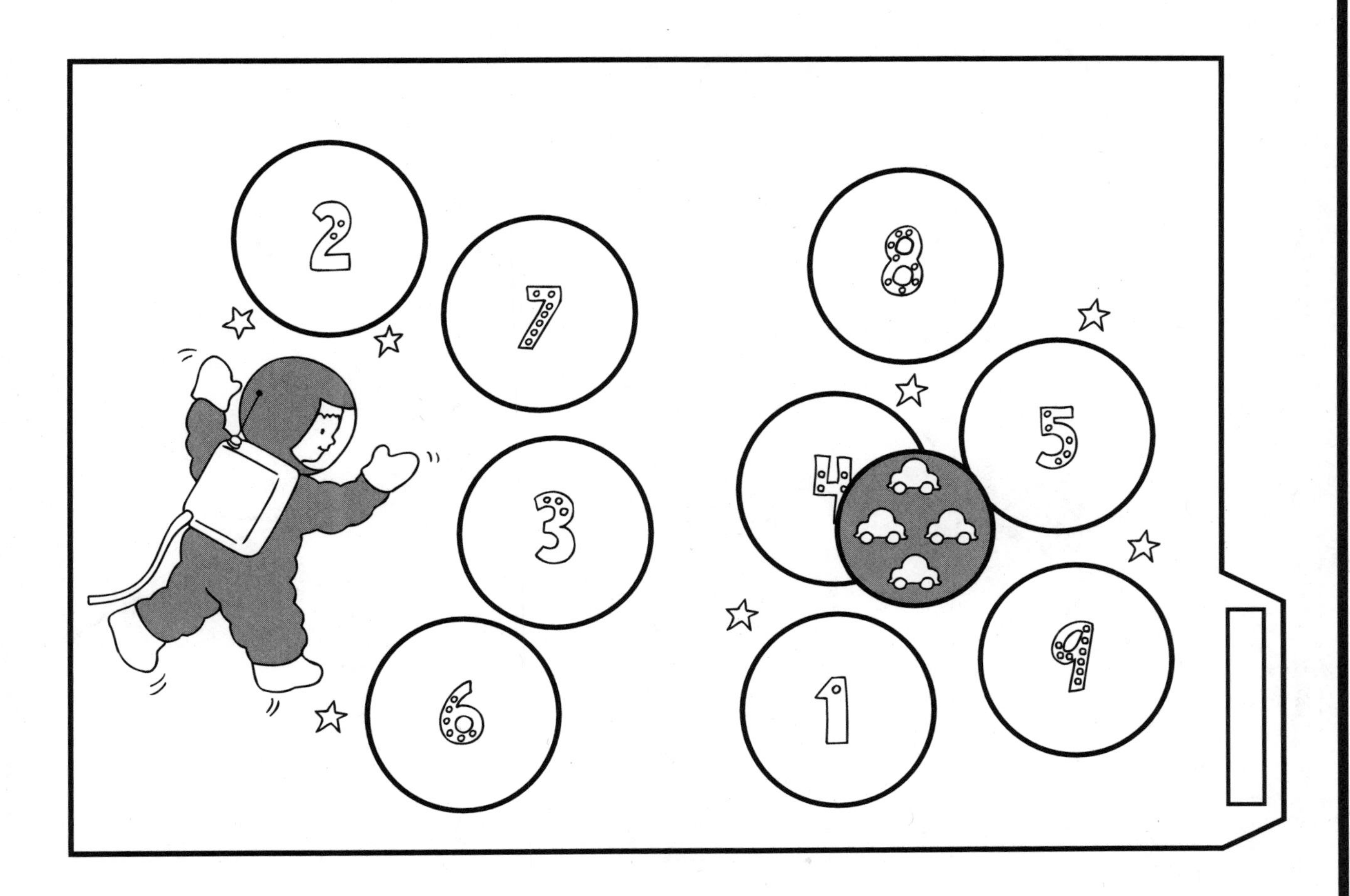

Space Walk

Counting/matching number symbols to sets

Directions: Two children may play. Take out the bubbles and open the folder. Put the bubbles face down on the playing area. Take turns turning over a bubble. Count the things in the bubble and name the number. Place the bubble on the matching number bubble on the Space Walk folder. To check, count the dots on the folder bubble.

A book to read: I Want to Be An Astronaut by Byron Barton

Space Walk

Space Walk

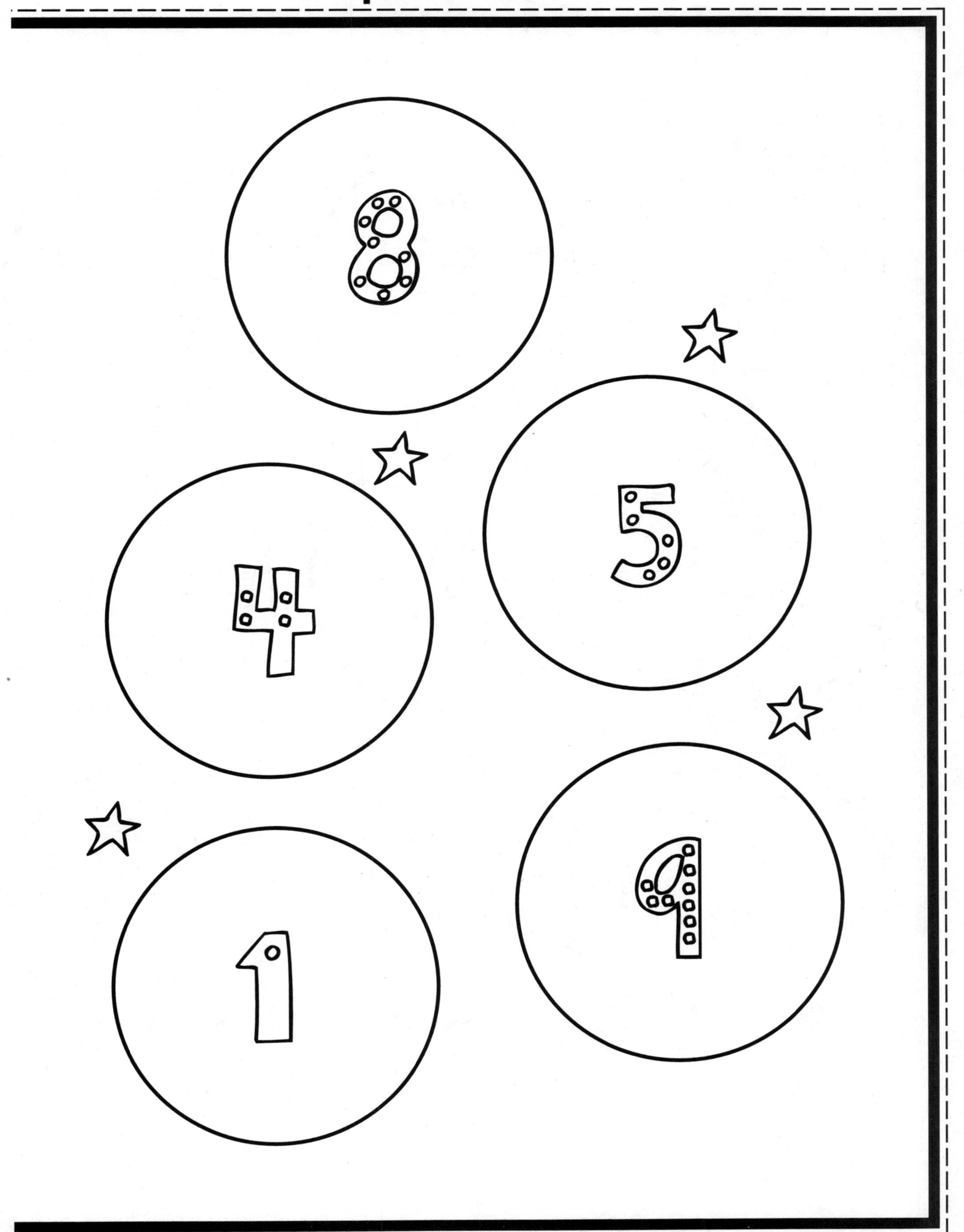

Space Walk

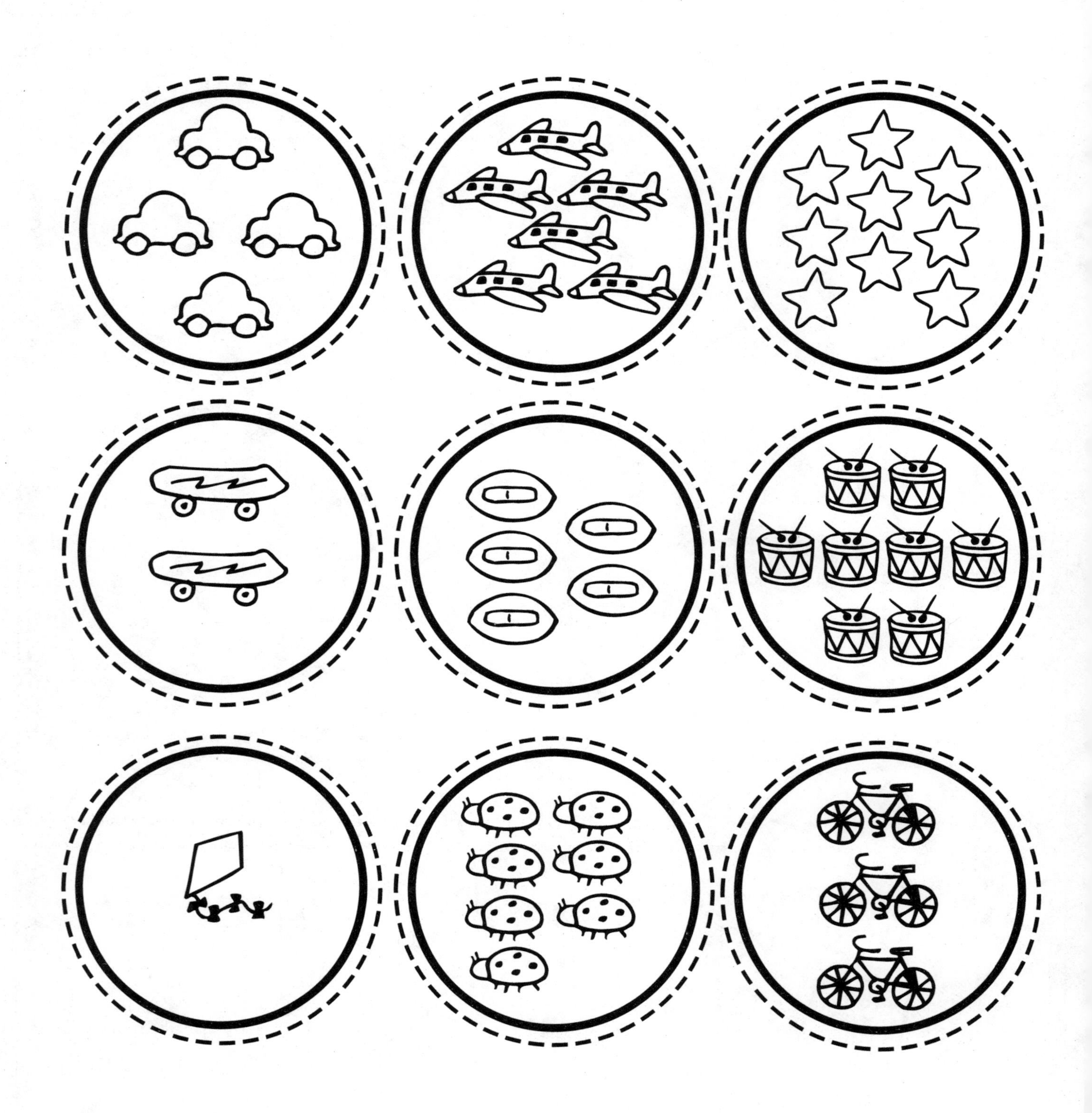

16. Balloons

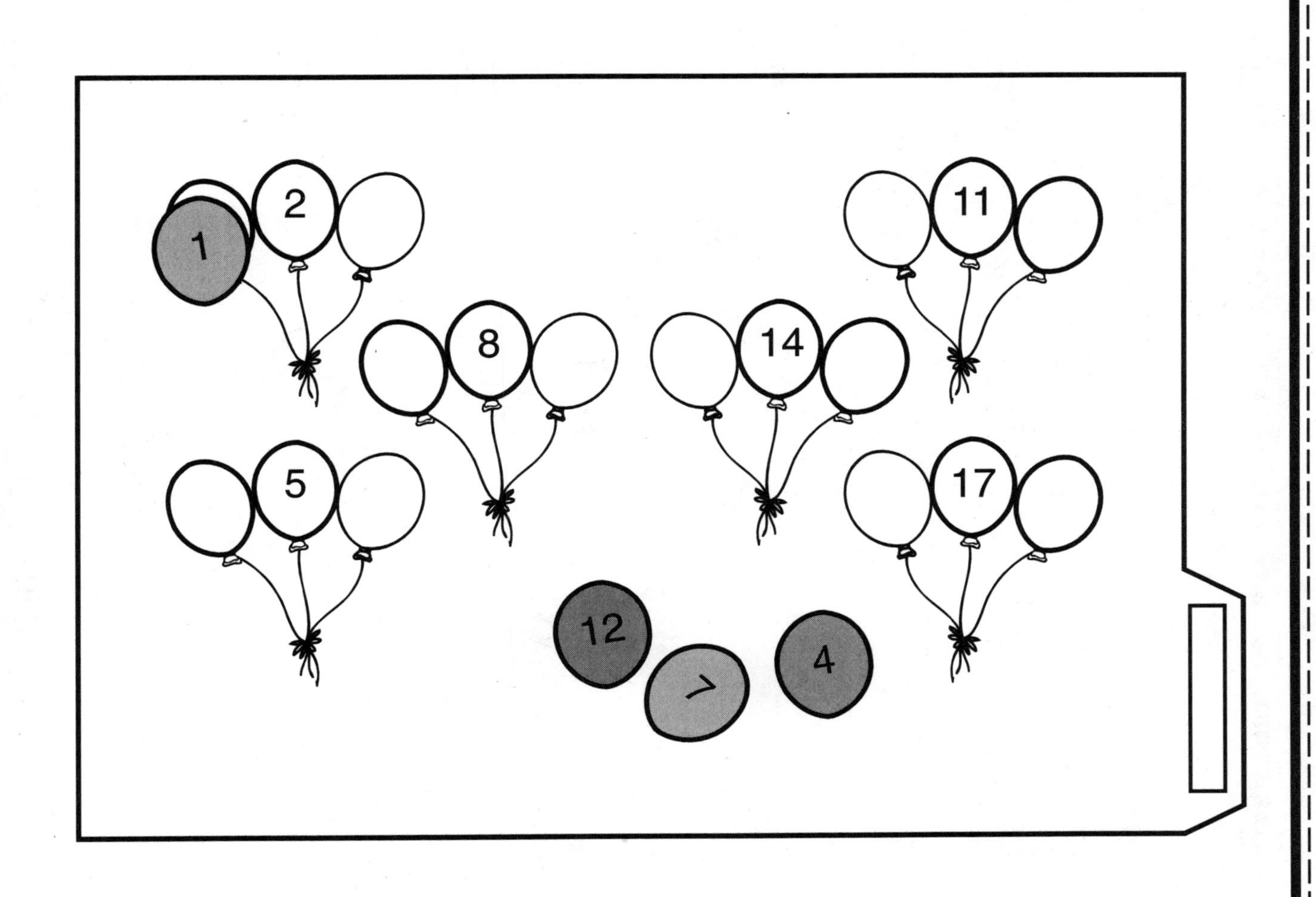

Balloons

Number sequence/before and after

Directions: Take out the numbered balloon circles and open the folder. Put the circles in a row face up. Find the missing number circles that go before and after the numbered balloon in each set. Put the number circles on the empty balloons. To check, name the numbers on the balloons in order.

A book to read: The Red Balloon by A. Lamorisse

Balloons

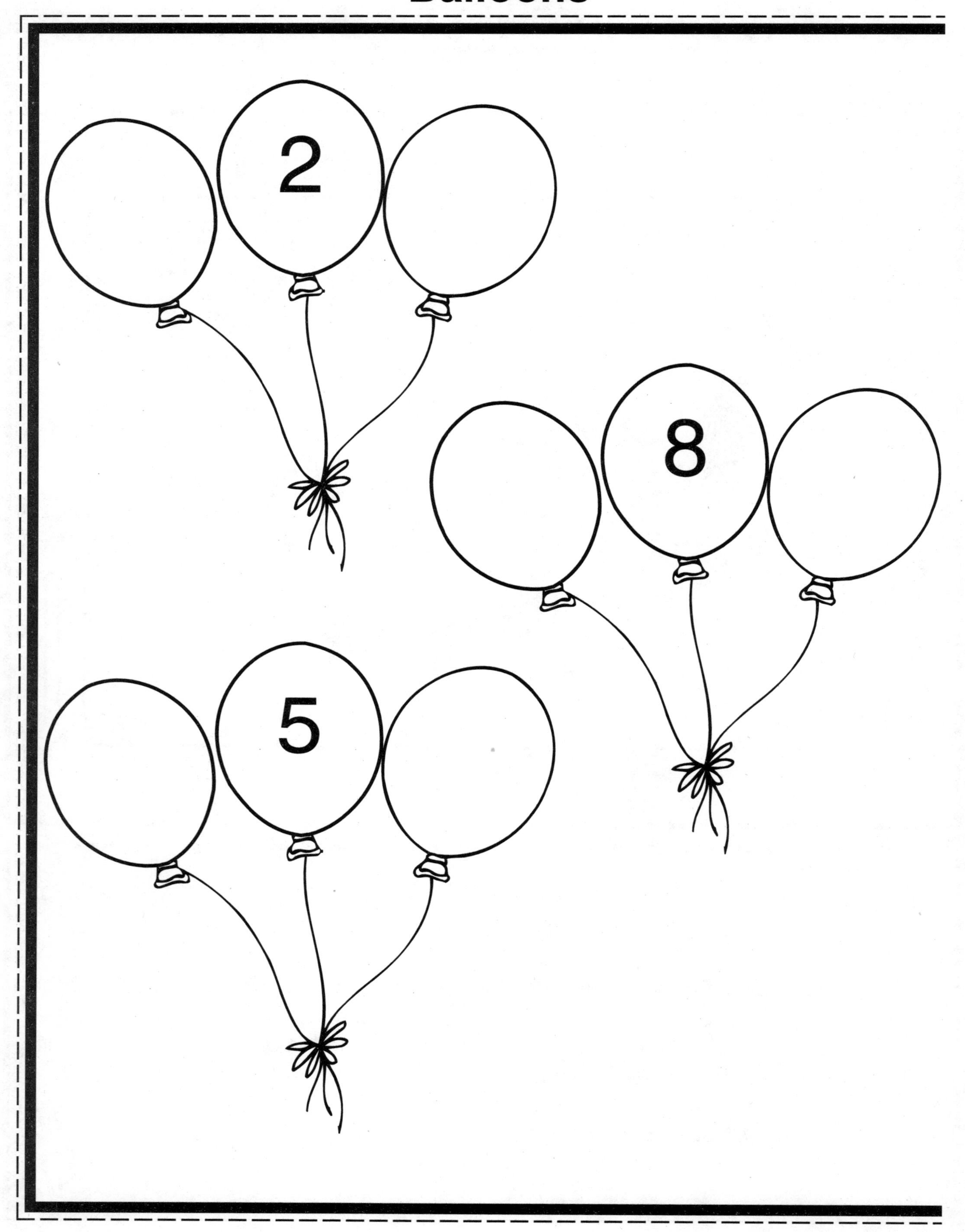

Balloons

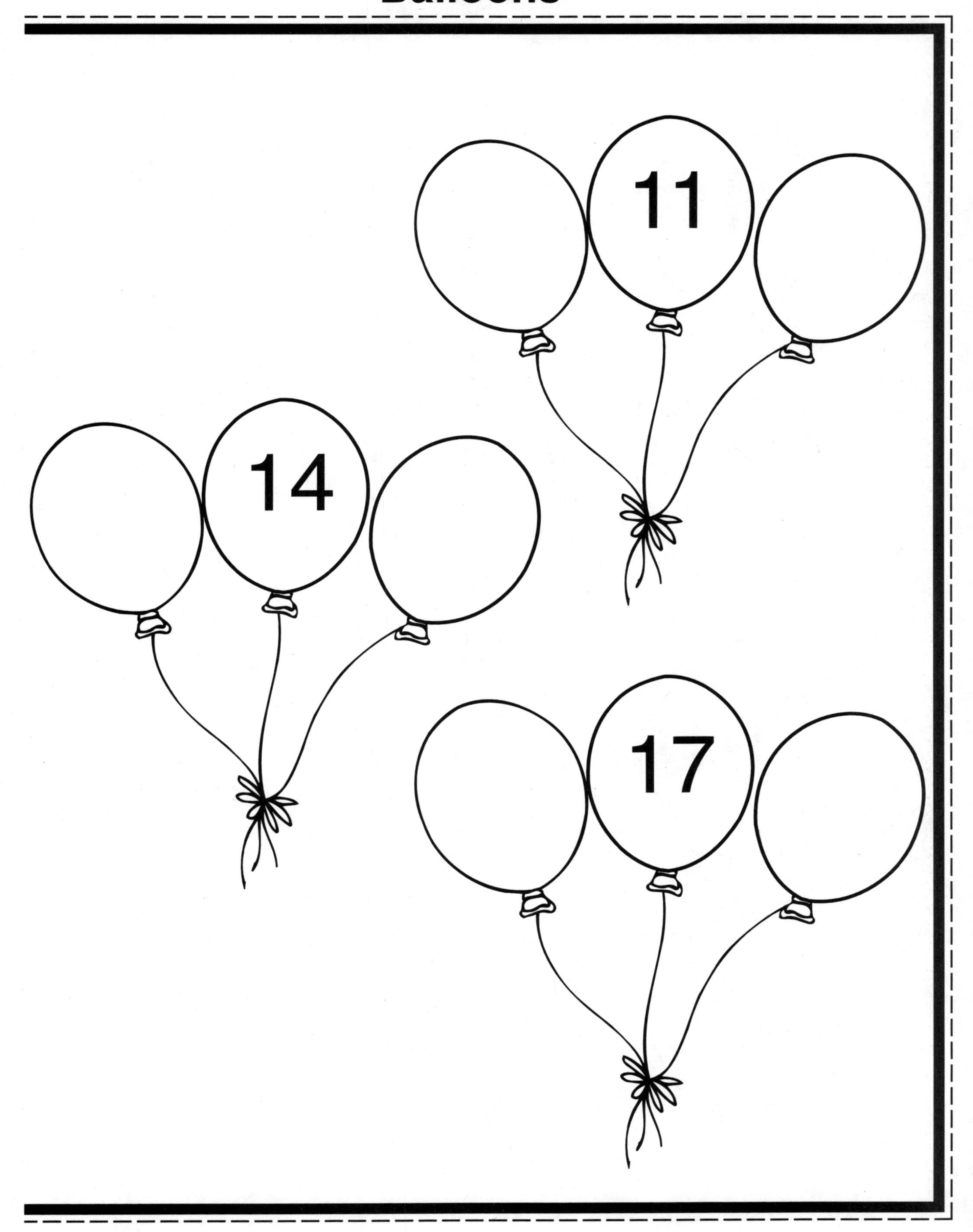

Balloons

17. Ice Cream Cones

Ice Cream Cones

Number sequence/number in between

Directions: Two children may play. Put the ice cream scoops in rows face down. Take turns picking a scoop. Find the scoop that goes between the two ice cream cones in each set. To check, read the numbers for each set of ice cream cones in order. For the activity card, use a washable felt pen or wipe-off crayon. Trace the number between the scoops in the ice cream sundaes. Read the numbers in order.

A book to read: Ice Cream Is Falling by Shigeo Watanabe

Ice Cream Cones

Ice Cream Cones

Ice Cream Cones

18. Tricky Clowns

Tricky Clowns

Beginning graphing

Directions: Take out the block cutouts and open the folder. Put the blocks in rows. See how many blocks it takes to fill each column for the tricky clowns. Which column takes the most blocks? Which one takes the fewest blocks? Compare!

A book to read: Jingle, the Christmas Clown by Tomie de Paola

Tricky Clowns

Tricky Clowns

Tricky Clowns

19. Grasshopper's Pogo Hop

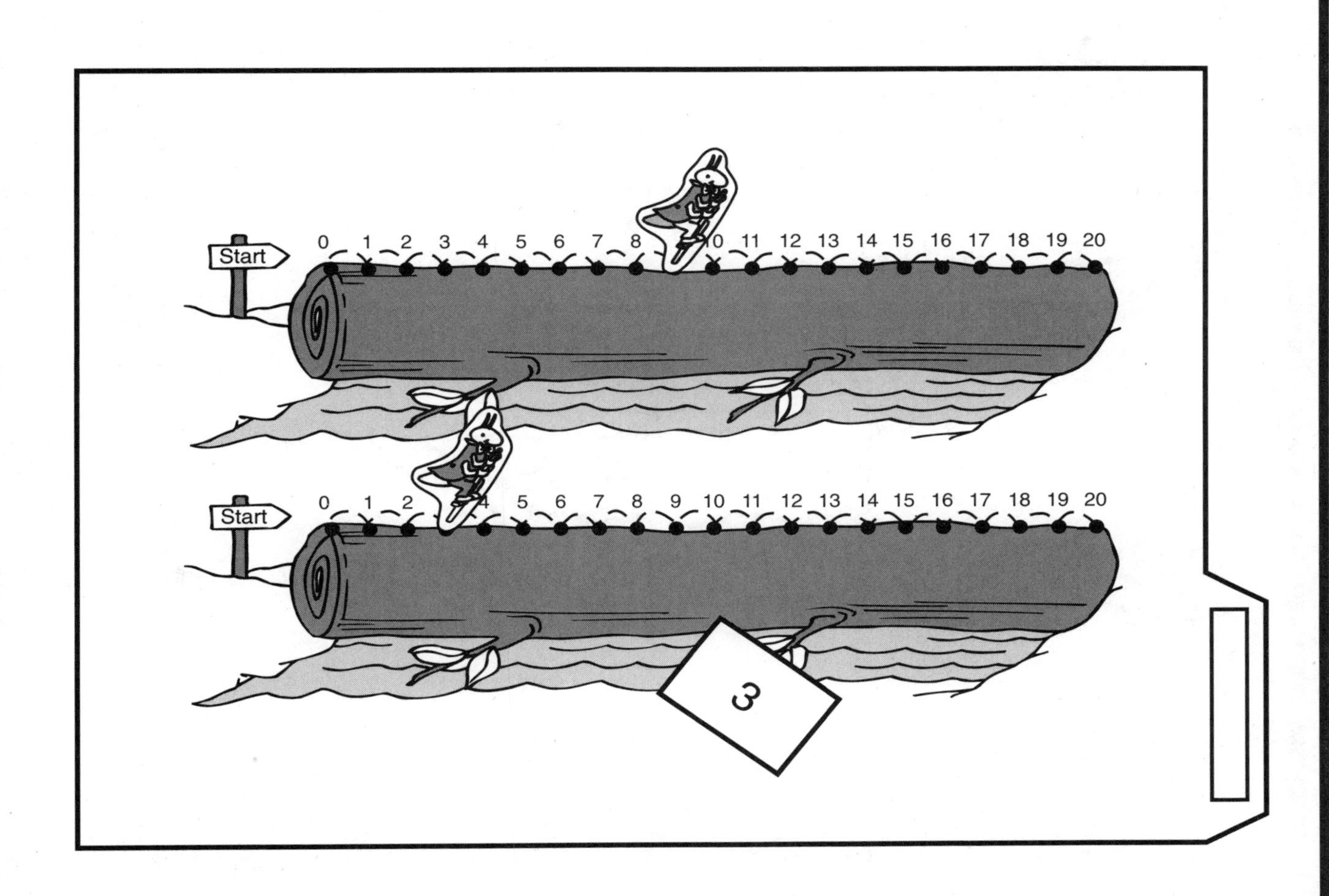

Grasshopper's Pogo Hop

Counting forward on number line, 0-20

Directions: Two children may play. Place the number cards face down. Choose a number log for your grasshopper to hop on. Put your marker at Start. In turn, pick a number card and hop your grasshopper forward that number of spaces. Count as you move forward. See who gets to the end of a log first. To be the winner, you must choose the number card that lets your grasshopper stop right at 20.

A book to read: Grasshopper on the Road by Arnold Lobel

Grasshopper's Pogo Hop

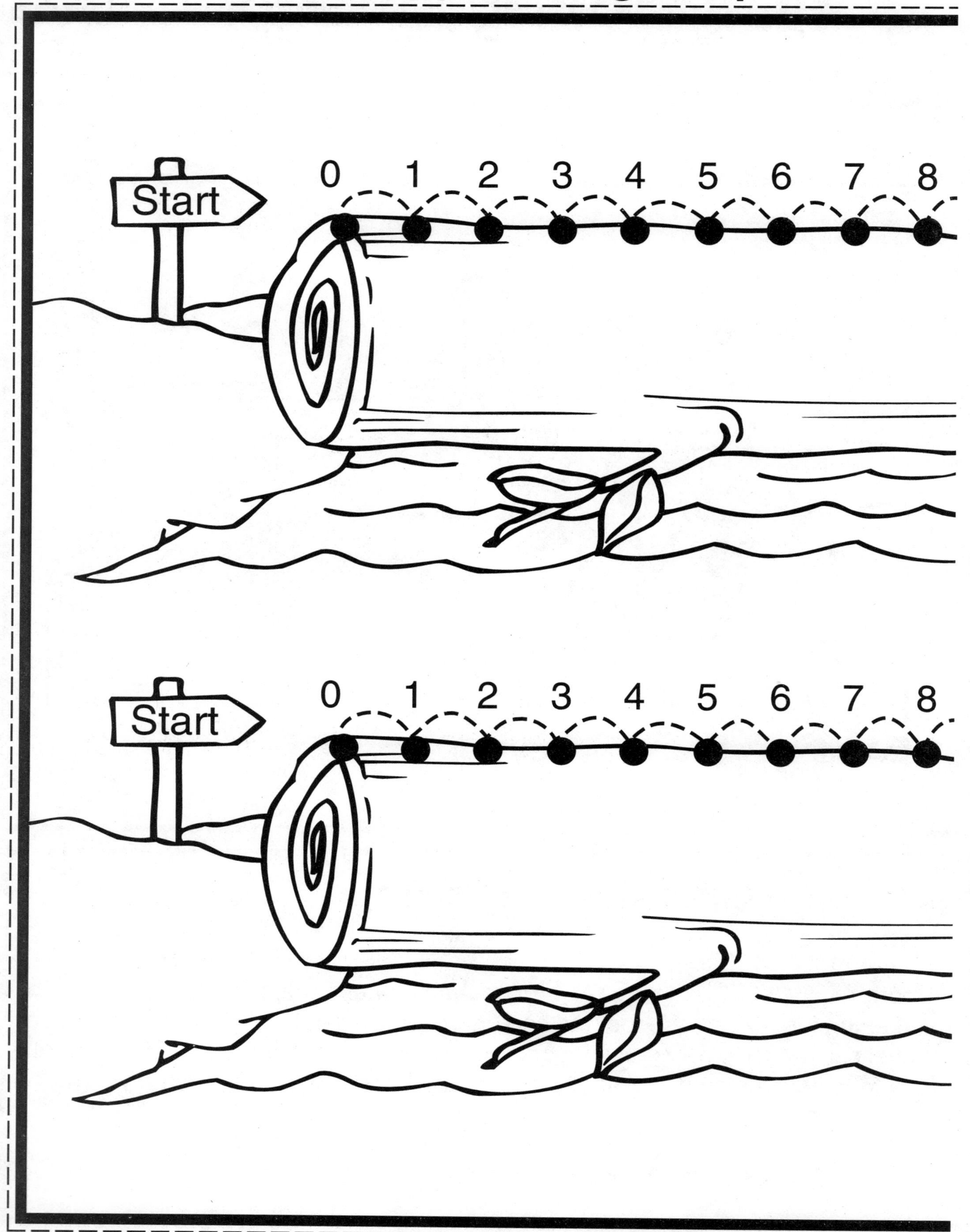

Grasshopper's Pogo Hop

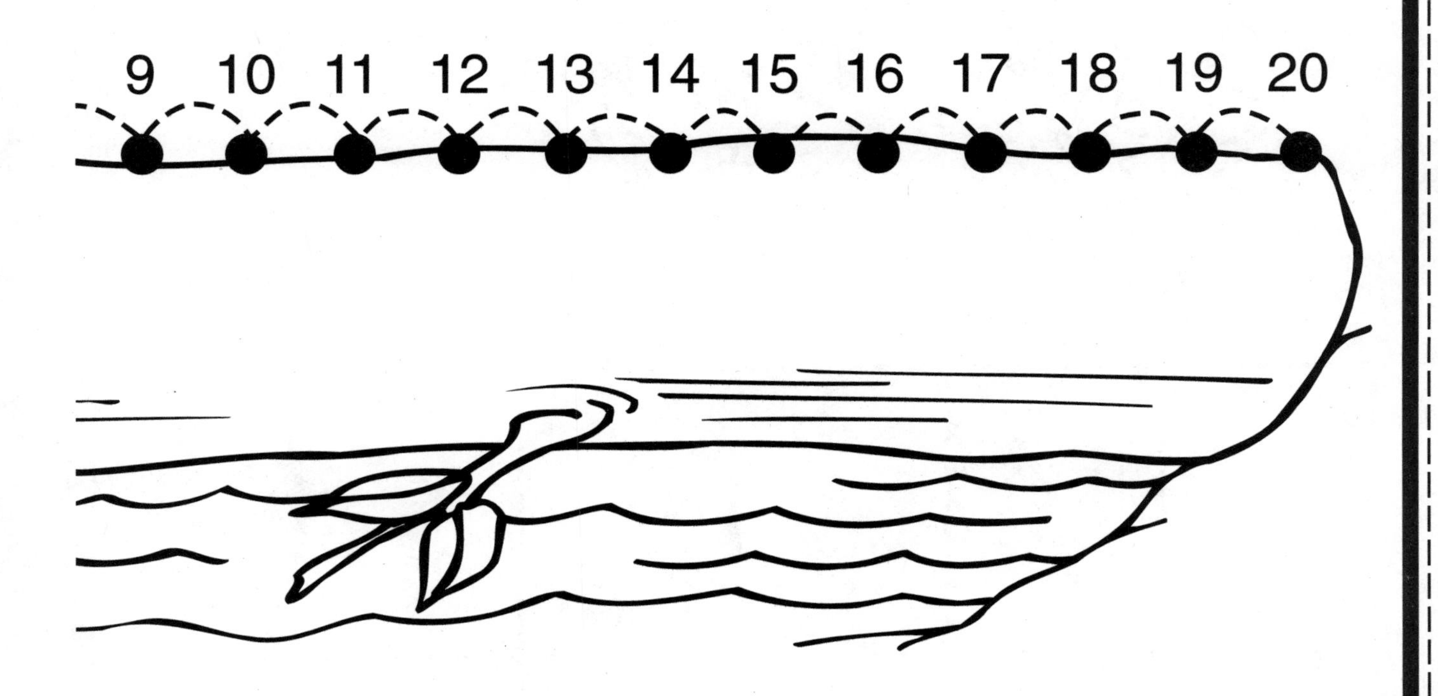

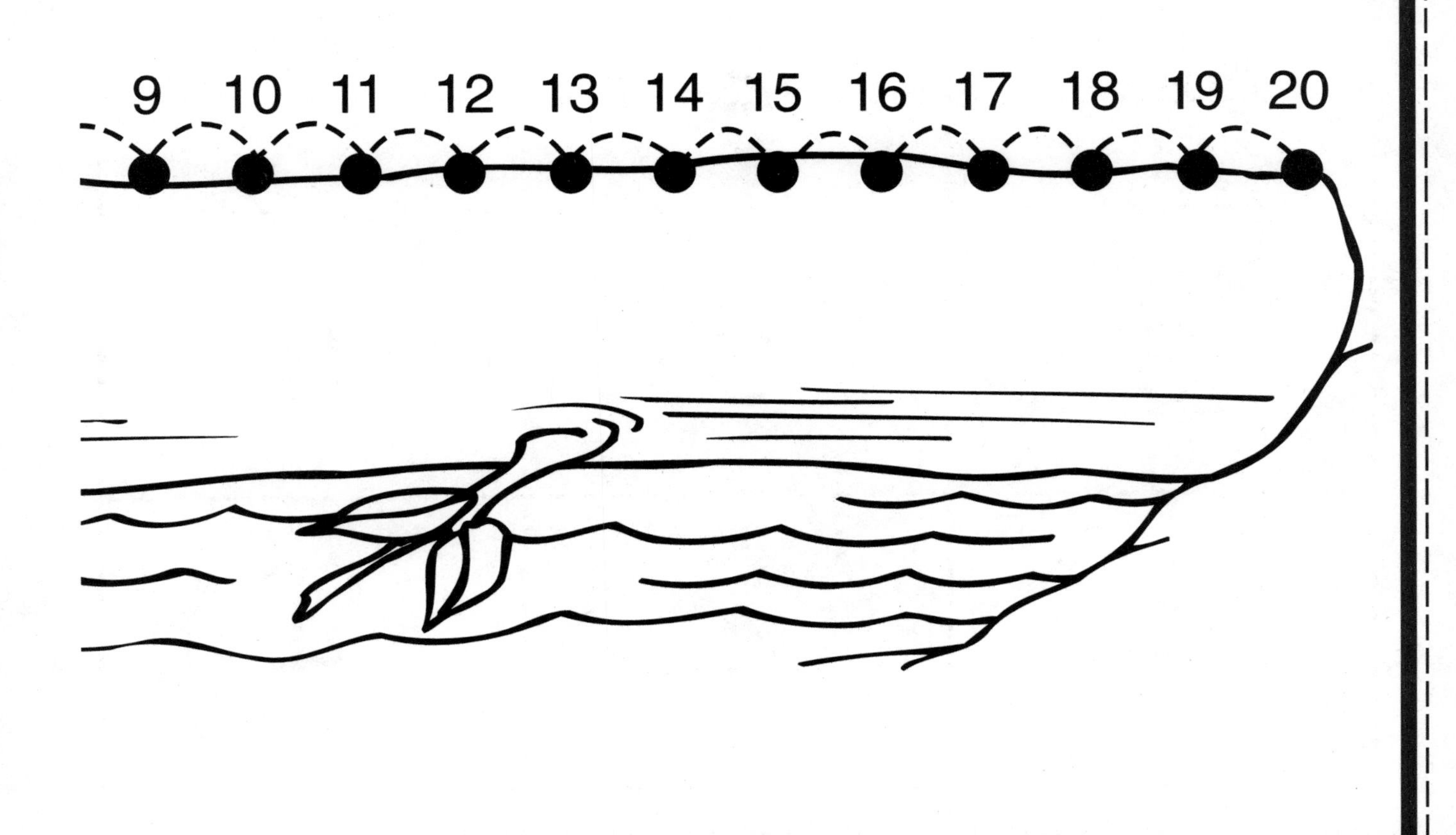

Grasshopper's Pogo Hop

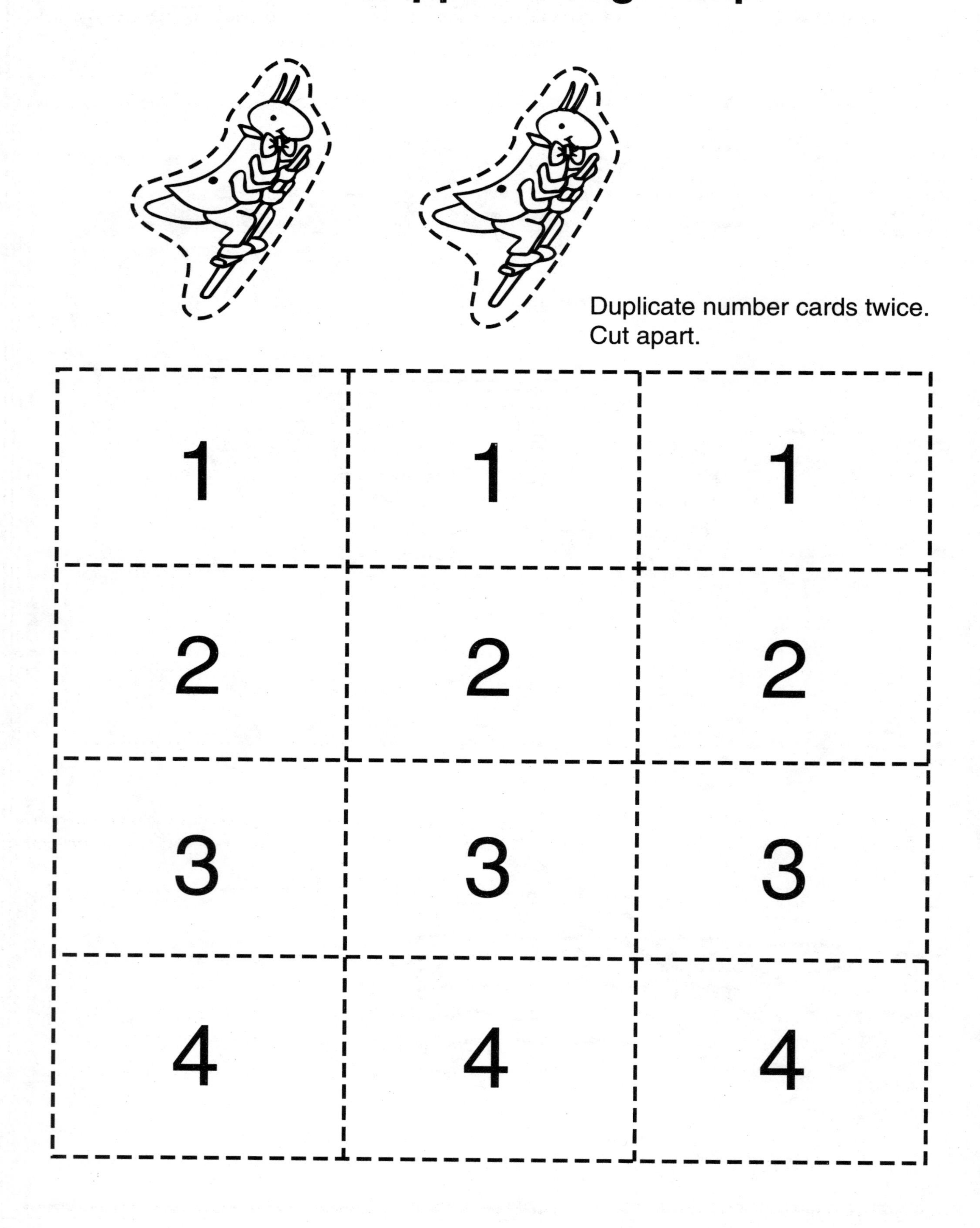

20. Mole's Stroll

Mole's Stroll

Counting backward on number line, 20-0

Directions: Two children may play. Take a mole marker and place the number cards face down. Put your mole on Start at the end of the tunnel, coming out. In turn, pick a number card and move your mole out of the tunnel that number of spaces. Count backward as you go. See which mole gets out of the tunnel first. To be the winner, you must choose the number card that lets your mole stop right at 0.

A book to read: <u>The Adventures of Mole and Troll</u> by Tony Johnson

Mole's Stroll

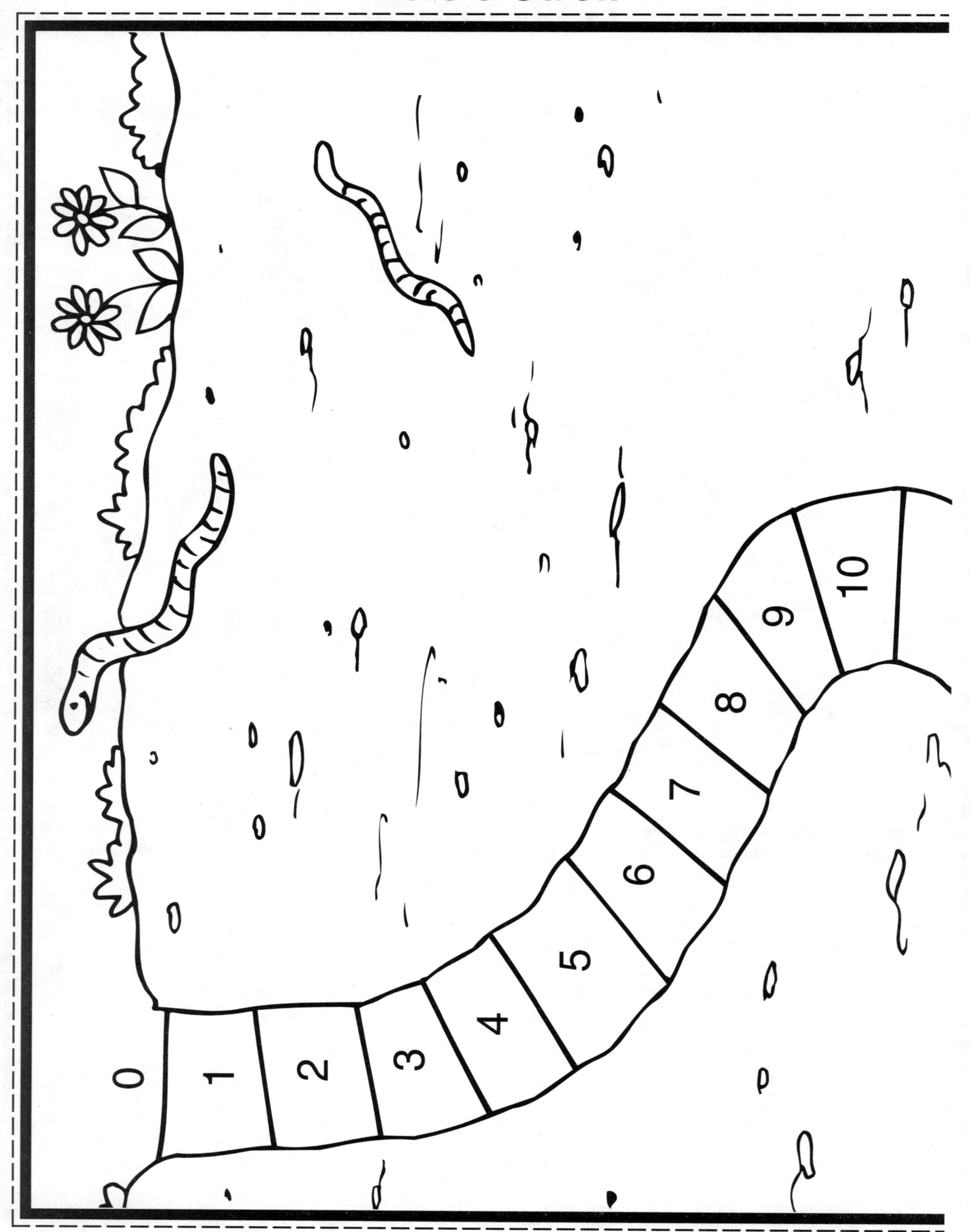

Mole's Stroll

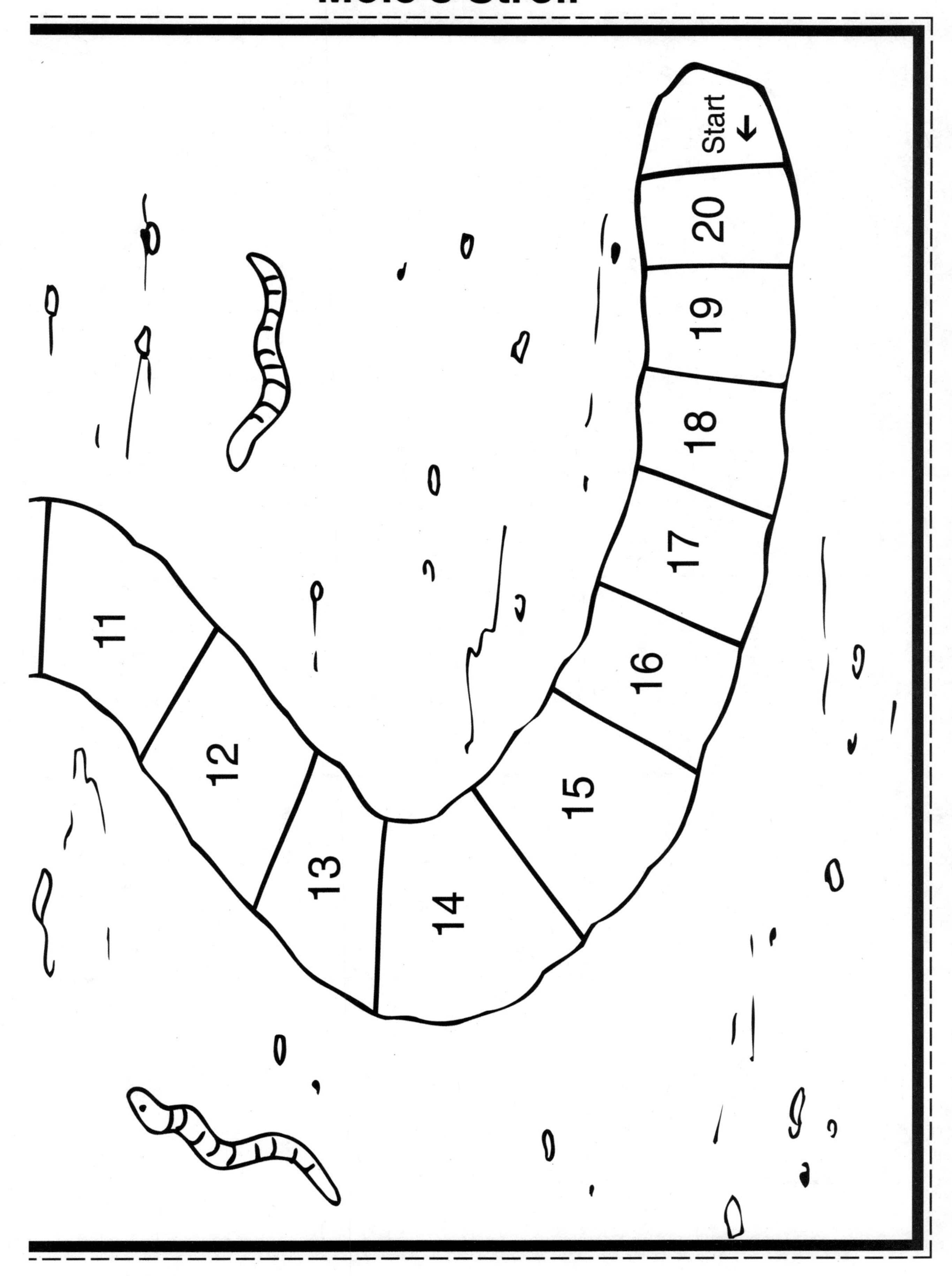

Mole's Stroll

Duplicate number cards twice.
Cut apart.

1	1	1
2	2	2
3	3	3
4	4	4

21. Ladybugs

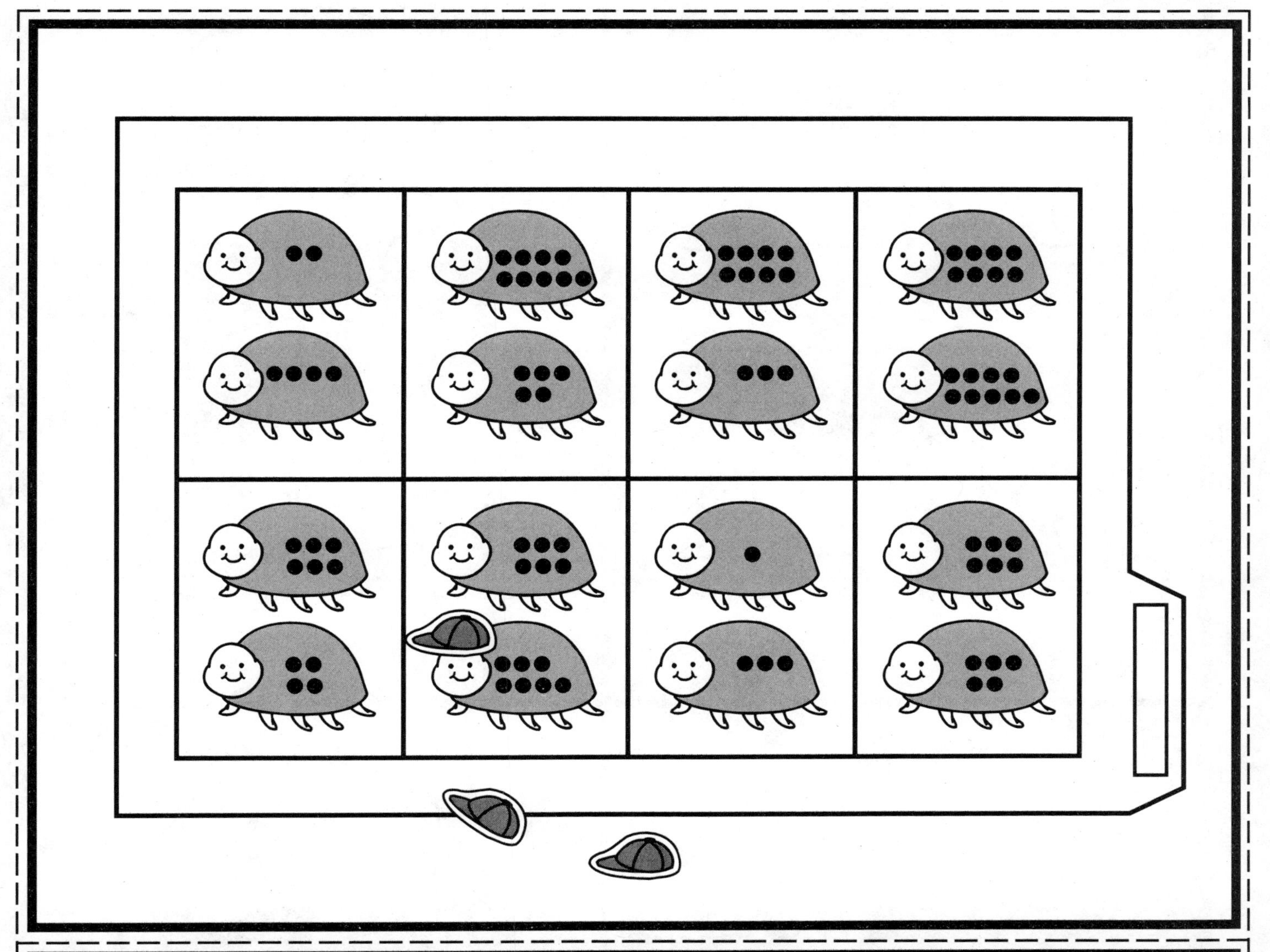

Ladybugs

Which is more?

Directions: Take out the ladybugs' caps and open the folder. Put the caps in a row. Count and tell how many dots there are on each ladybug in each set. Put a cap on each ladybug that has more dots than the other.

A book to read: The Grouchy Ladybug by Eric Carle

Ladybugs

Ladybugs

Ladybugs

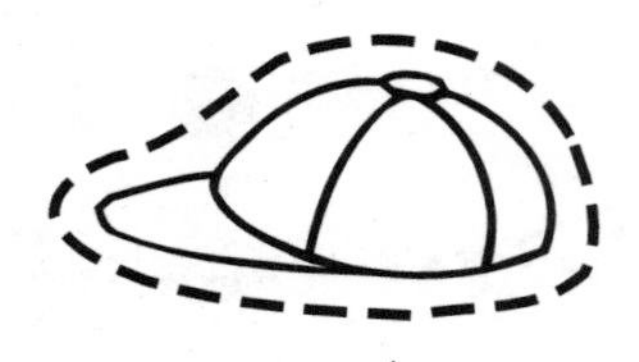

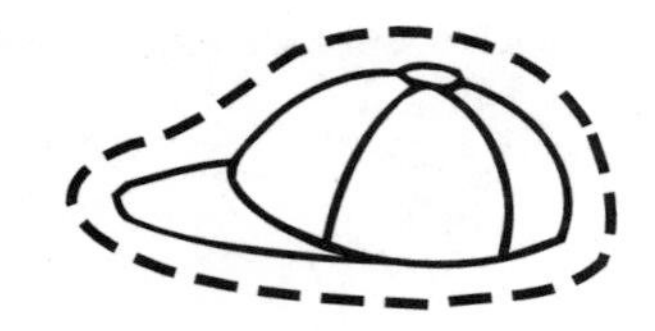

22. Apples and Worms

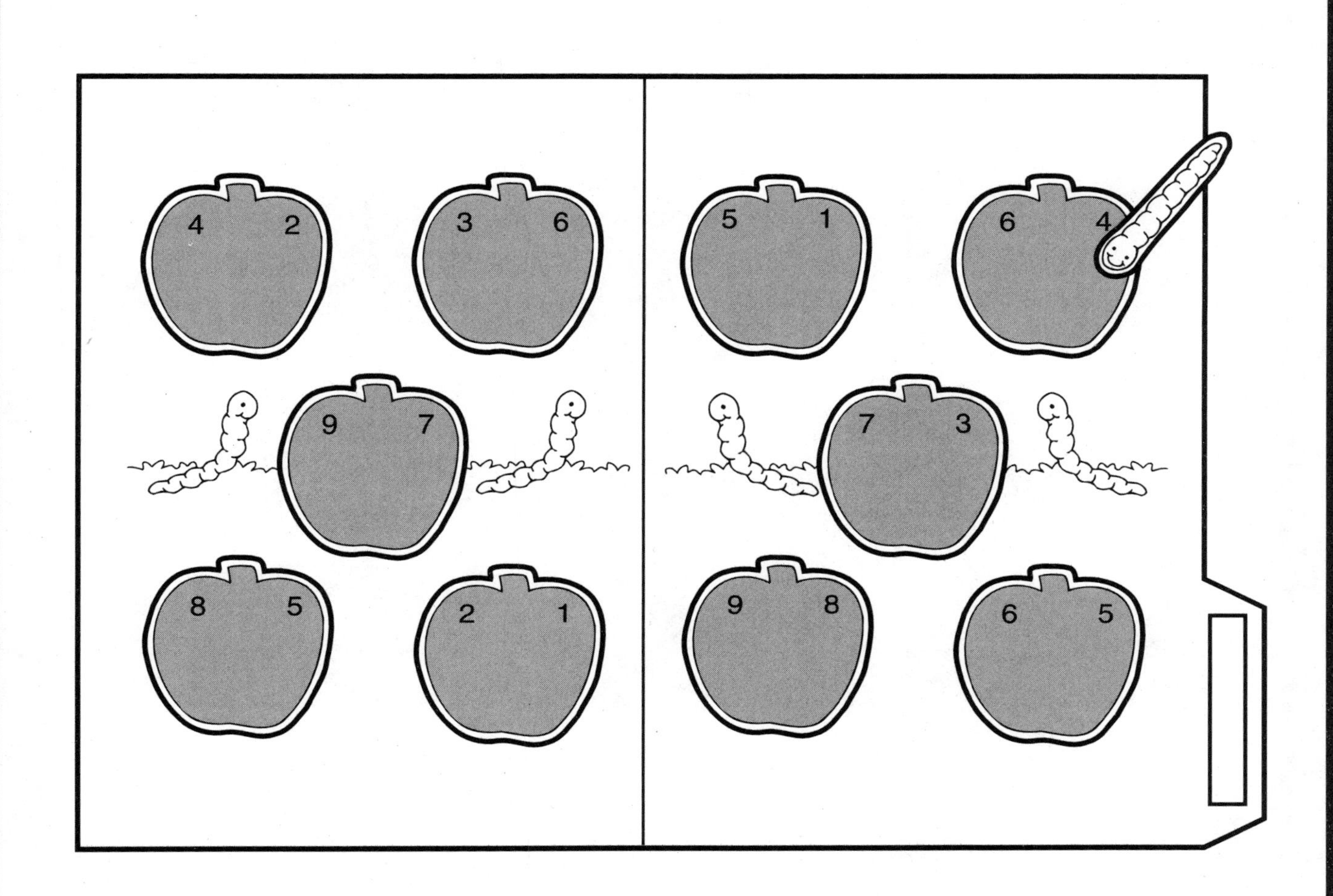

Apples and Worms

Which is less?

Directions: Take out the worm clothespins. Open the folder. Look at the two numbers on each apple. Which one is less than the other? Clip the worm on the number that is less on each apple.

A book to read: The Big Fat Worm by Nancy Van Laan

Apples and Worms

Apples and Worms

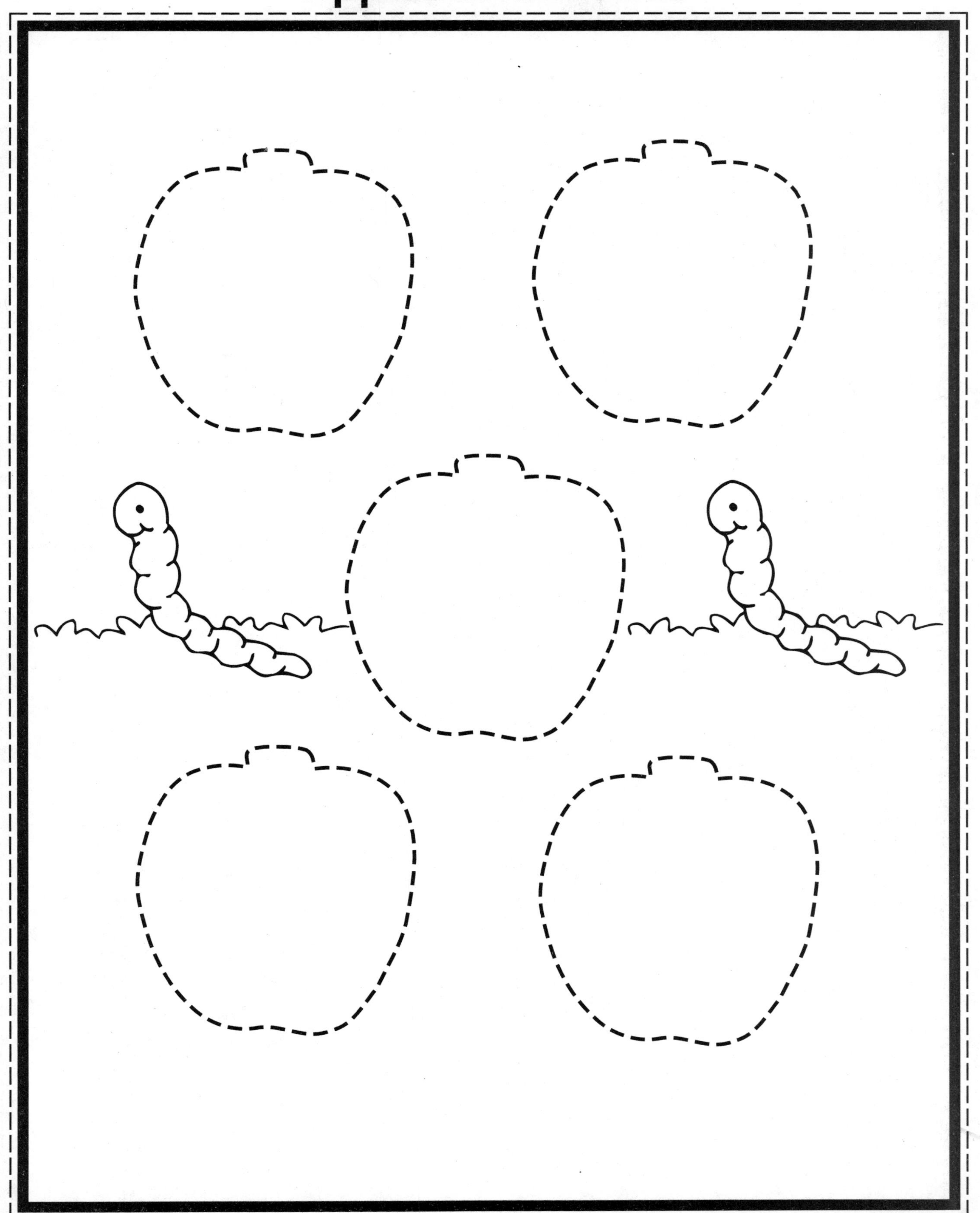

Apples and Worms

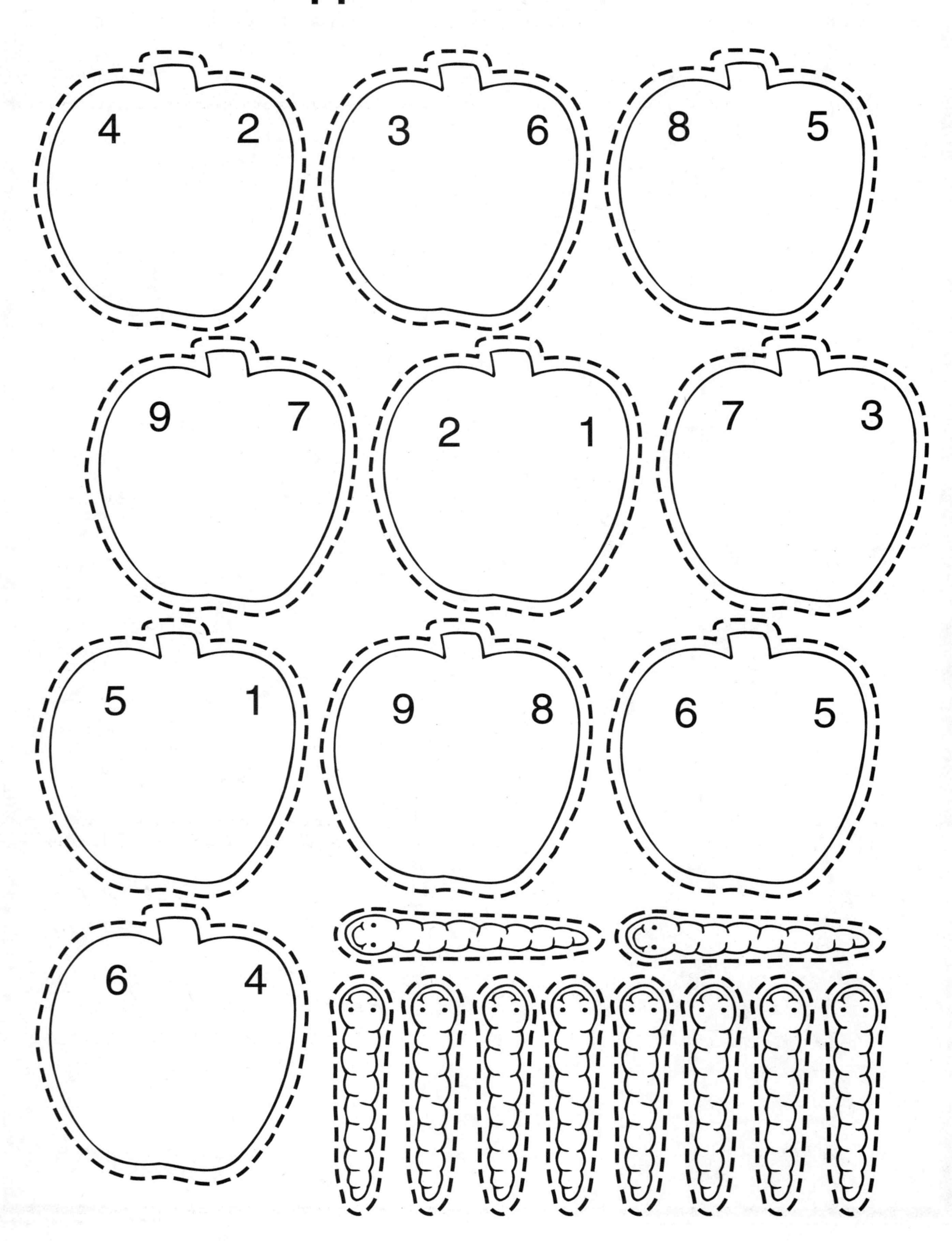

23. Muffins

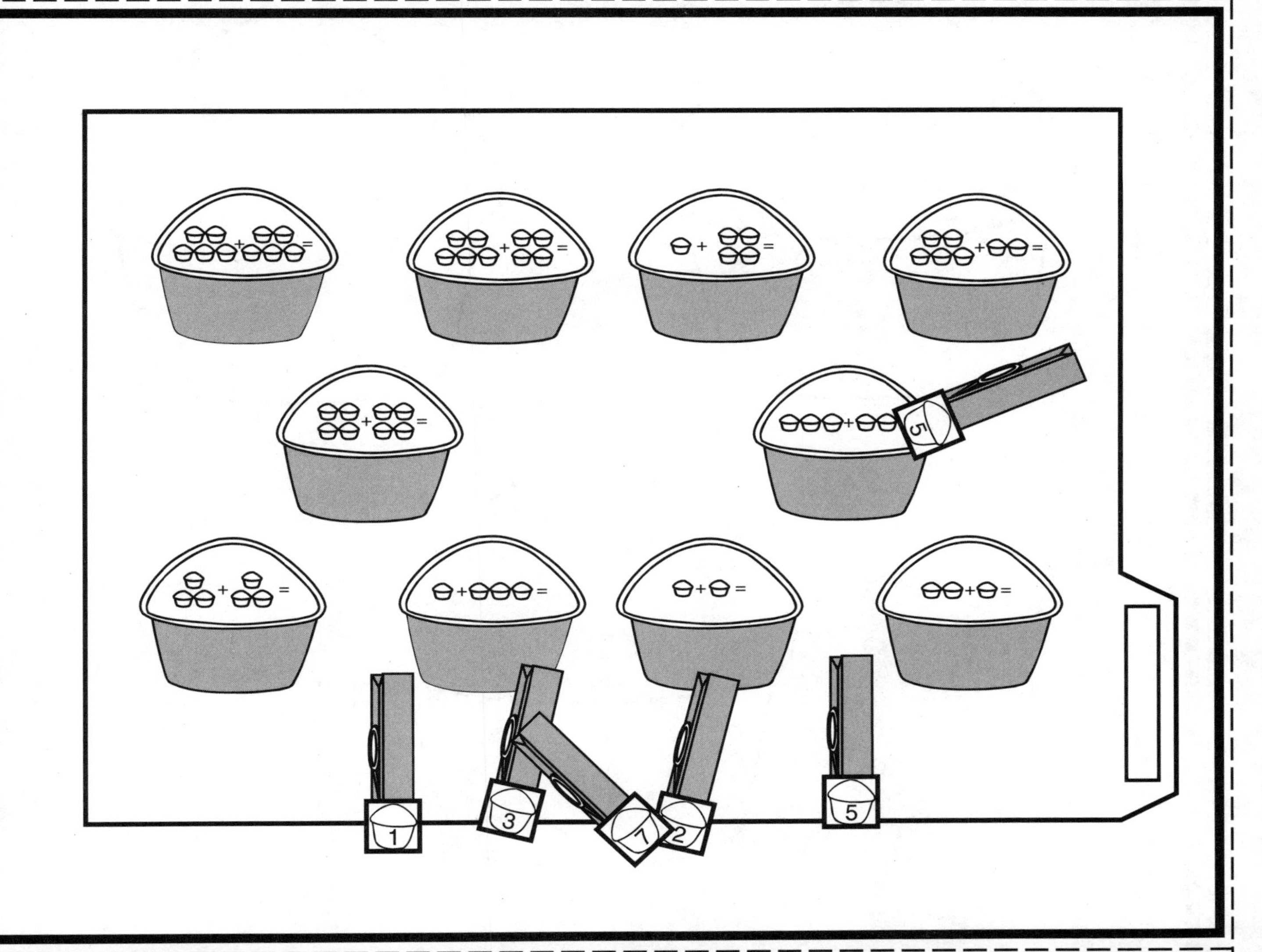

Muffins

Recognizing the + sign and the = sign

Directions: Take out the numbered muffin clothespins and open the folder. Look at the little muffins on each big muffin. The **+** sign tells you to add or put together. The **=** sign says equal. Count the little muffins on each big muffin and find out how many there are altogether. Then find the clothespin number that matches. Clip it on the big muffin after the **=** sign.

A book to read: If You Give a Moose a Muffin by Laura Jaffe Numeroff

Muffins

Muffins

Muffins

24. Nuts and Squirrels

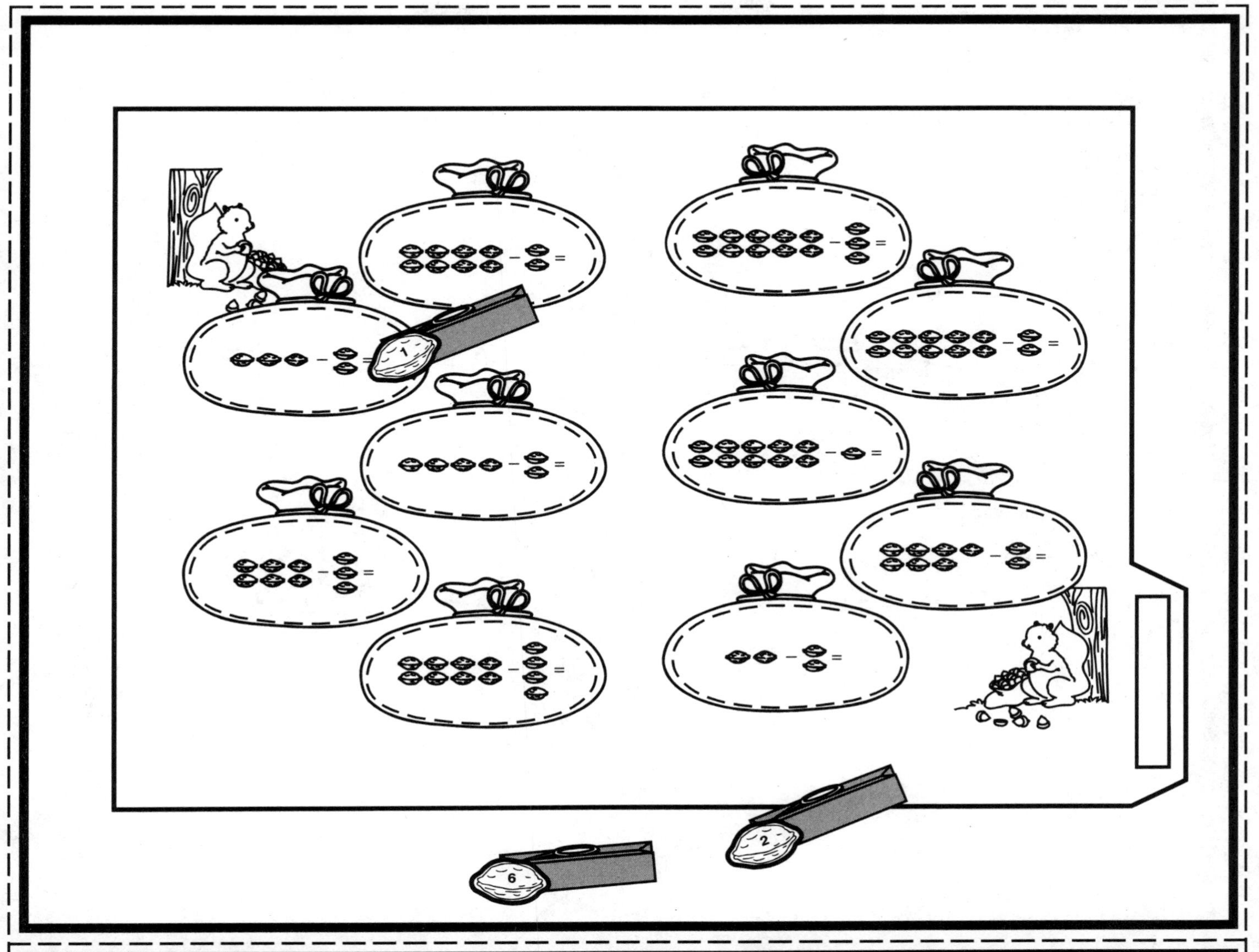

Nuts and Squirrels

Recognizing the - sign and the = sign

Directions: Take out the numbered nut clothespins and open the folder. Look at the squirrels' nut bags. The **-** sign means take away. The **=** sign says equal. Count the nuts in the big group on each bag. Count how many to take away. How many nuts are left? Find the numbered clothespin that matches how many nuts are left. Clip it on the bag after the **=** sign. Read your number story sentence. Example: Three nuts take away two nuts leaves one nut.

A book to read: Squirrels by Brian Wildsmith

Nuts and Squirrels

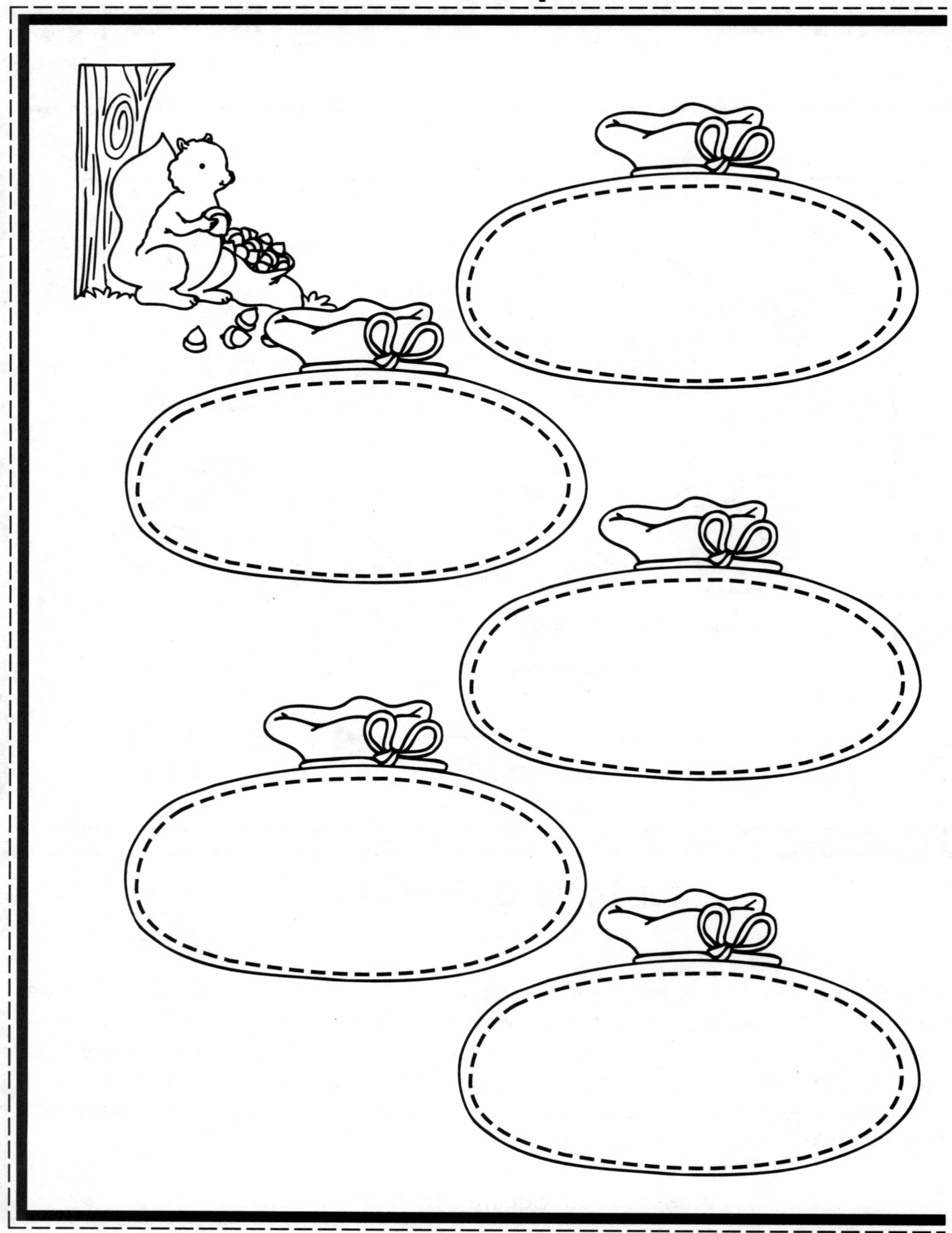

Nuts and Squirrels

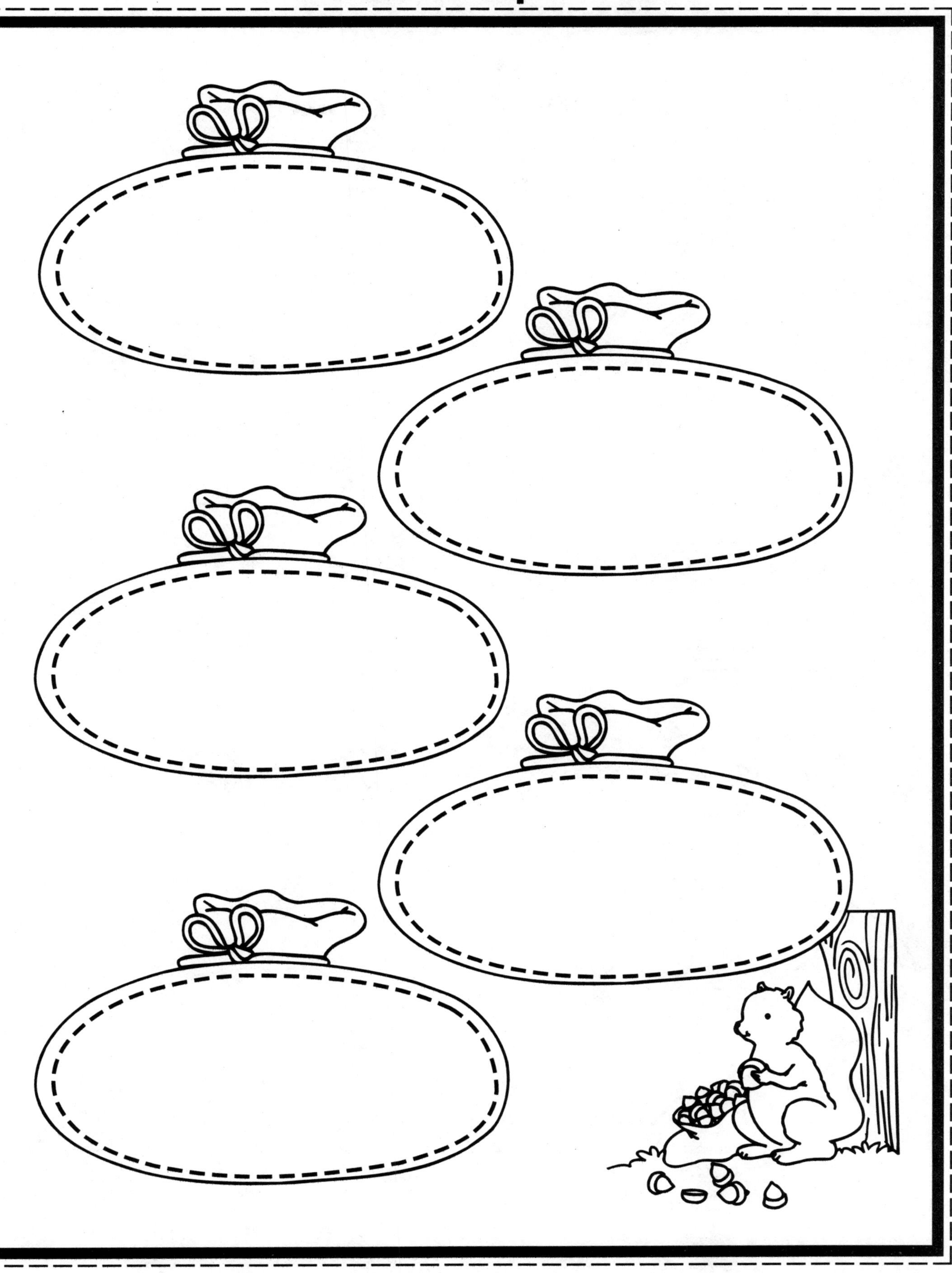

Nuts and Squirrels

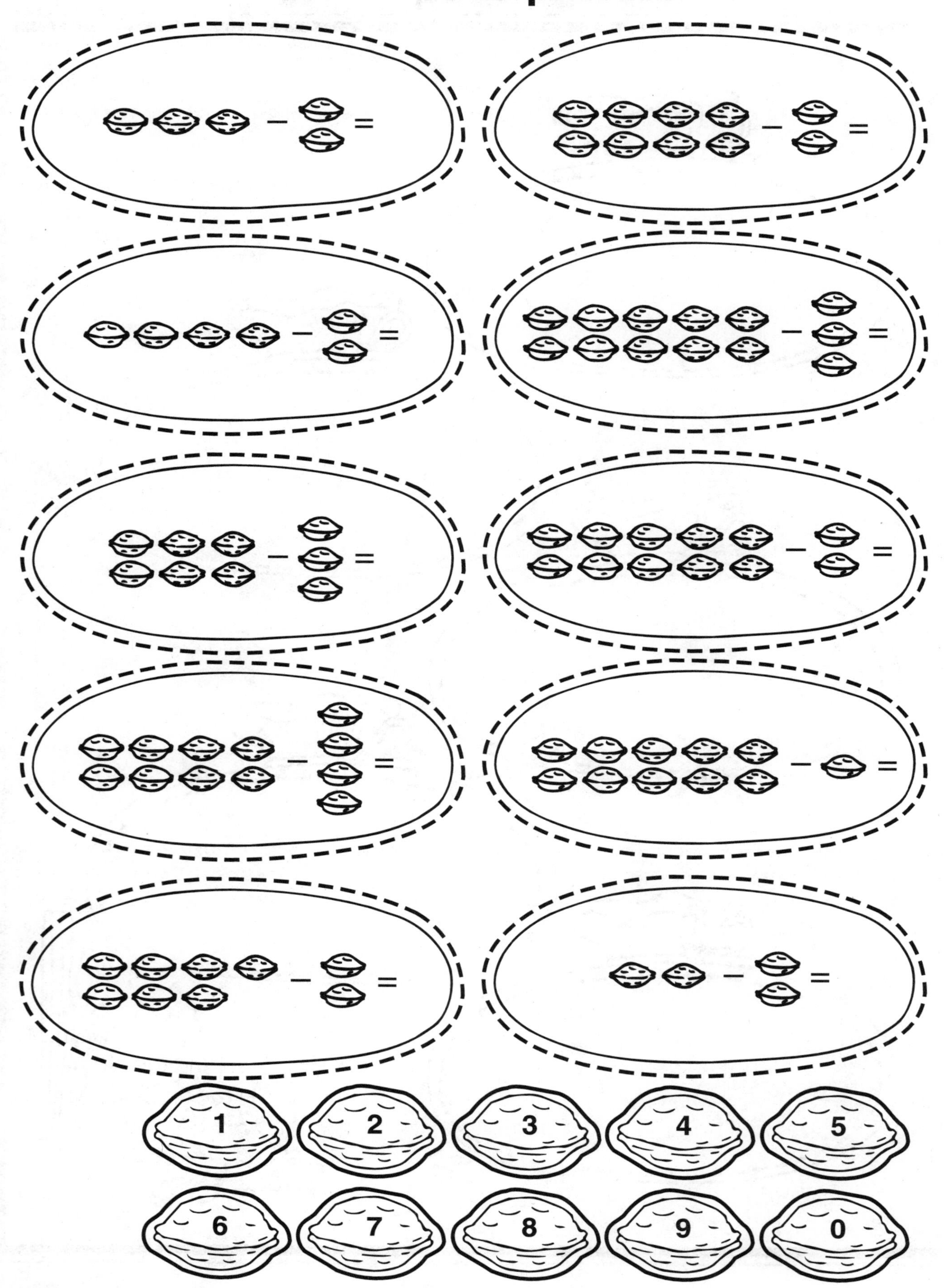

25. Rabbit Families

Rabbit Families

Different names for sets

Directions: Take out the clothespins and open the folder. Look at the rabbit families in each circle. Count and name the two sets in each family circle. Find the clothespin that matches the number name for the complete family. Clip it to the circle. Example: A set of two rabbits and three rabbits makes five rabbits altogether.

A book to read: The Runaway Bunny by Margaret Wise Brown

Rabbit Families

Rabbit Families

Rabbit Families

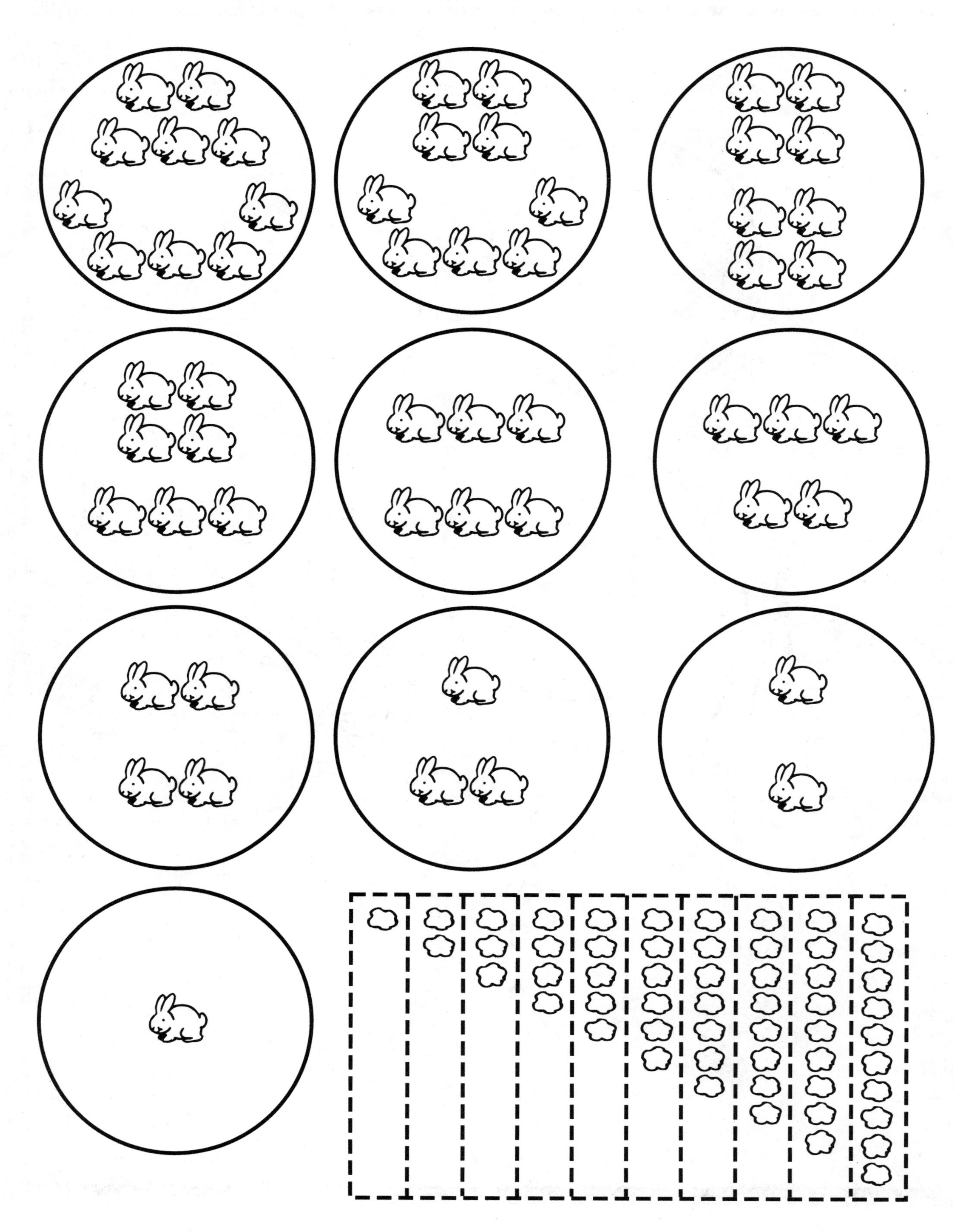

26. Pumpkin Patch

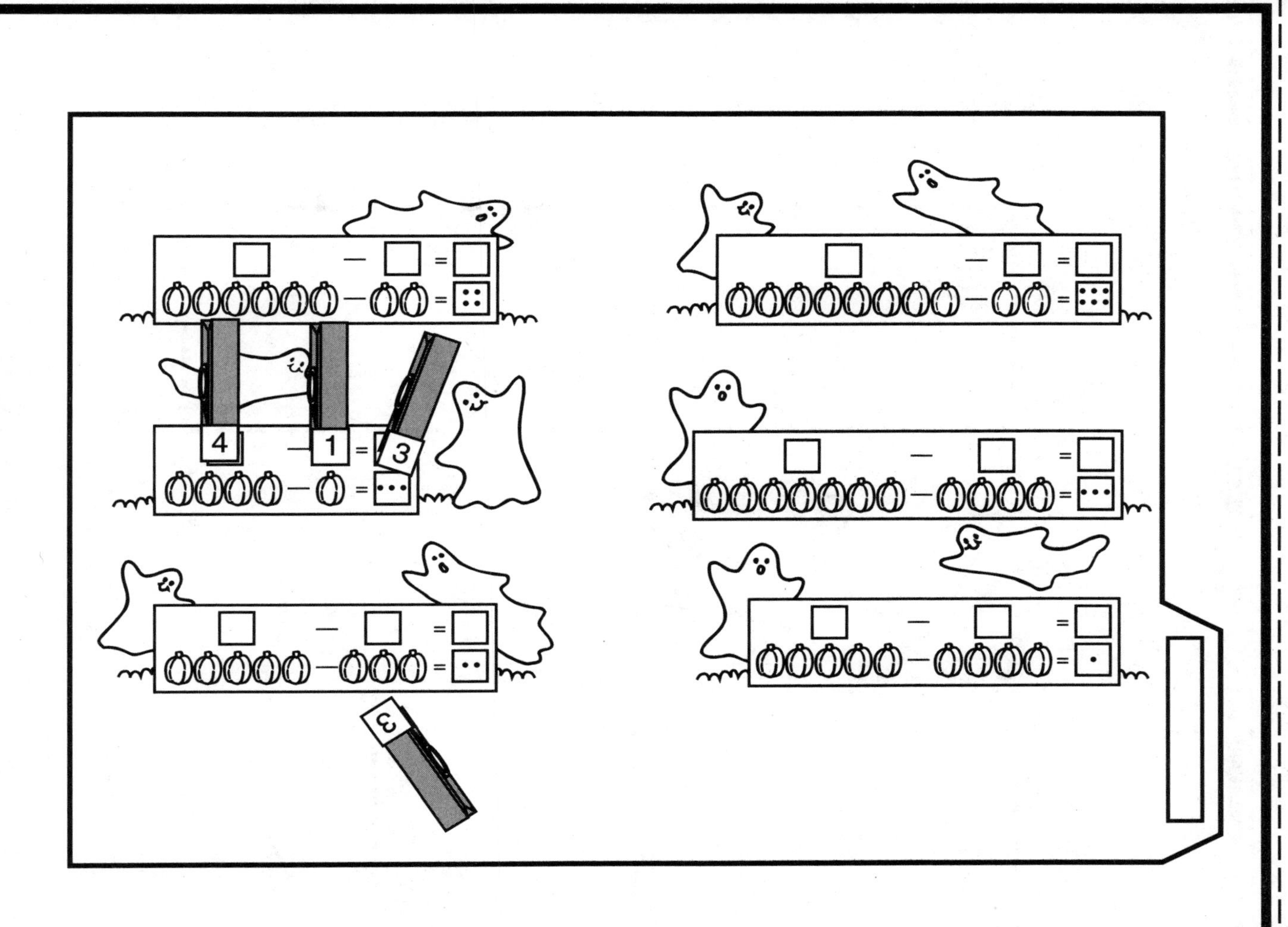

Pumpkin Patch

Subtraction number sequence with symbols

Directions: Look at the pumpkins in each pumpkin patch. The **-** sign says to take away. The = sign says equal. Count the big group of pumpkins on each fence. Clip the numbered clothespin that matches to the empty box. Count how many to take away. Clip the numbered clothespin that matches to the empty box. How many pumpkins are left? Clip the numbered clothespin that matches to the empty box. Check with the dots in the end box. Read your number story sentence. Example: Four pumpkins take away one pumpkin leaves three pumpkins.

A book to read: Jeb Scarecrow's Pumpkin Patch by Jana Dillon

Pumpkin Patch

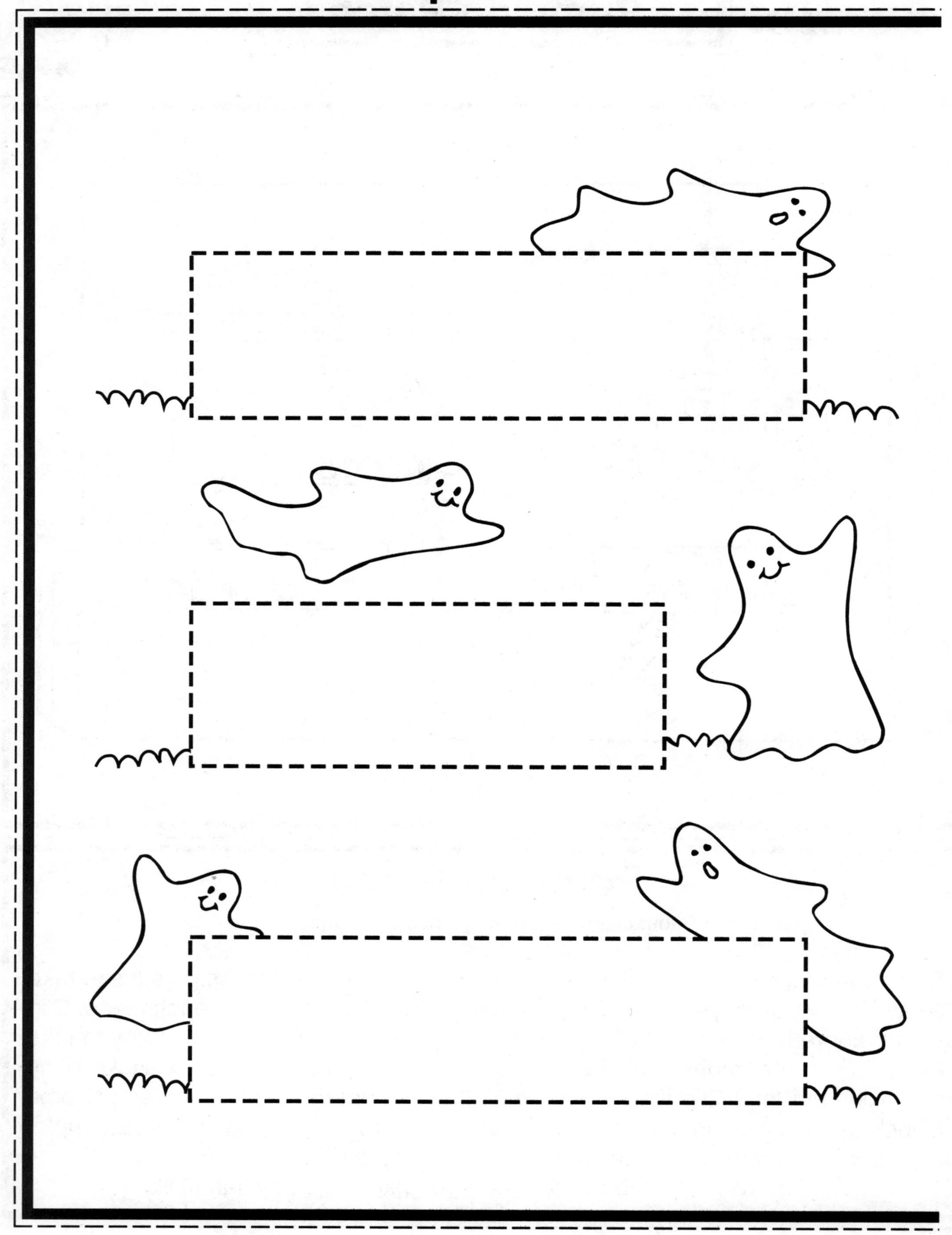

Pumpkin Patch

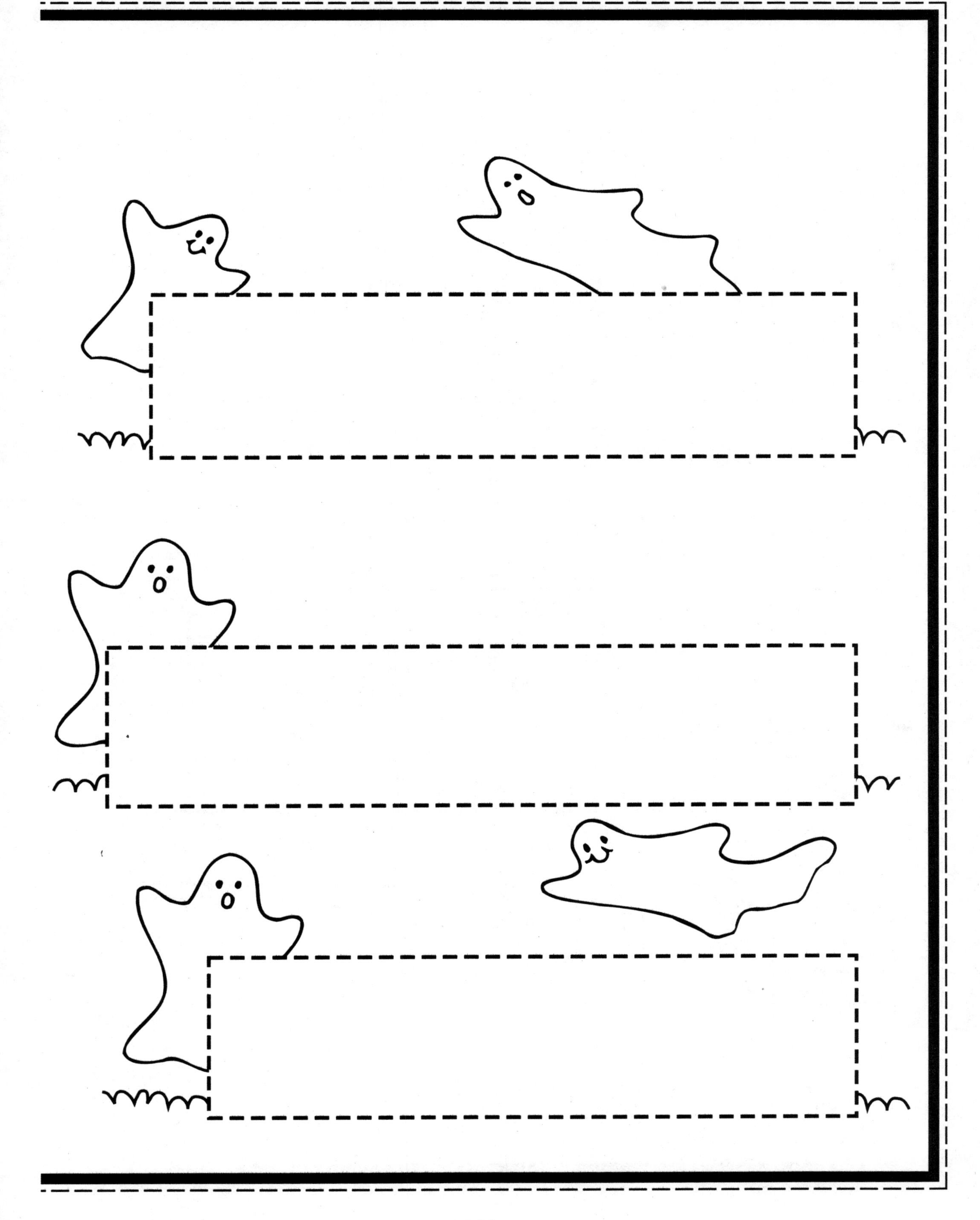

Pumpkin Patch

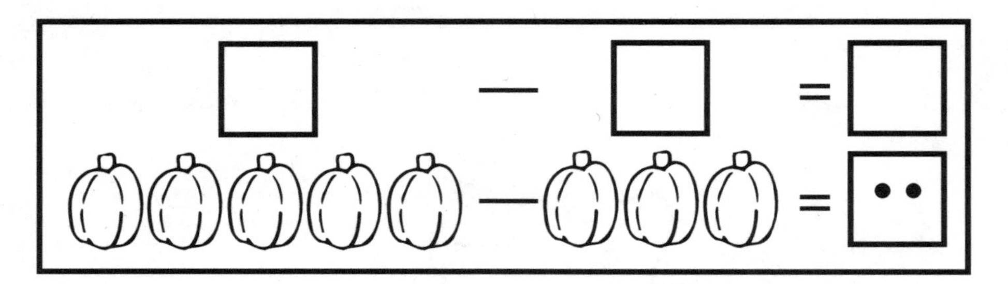

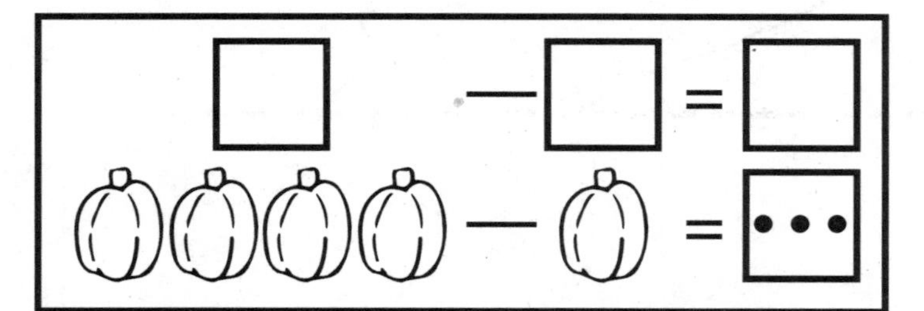

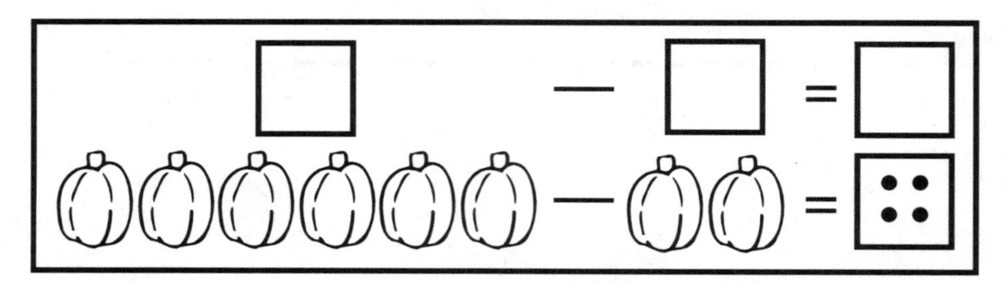

5
3
2
4
1
3
6
2
4
8
2
6
7
4
3
5
4
1

27. Card Mates

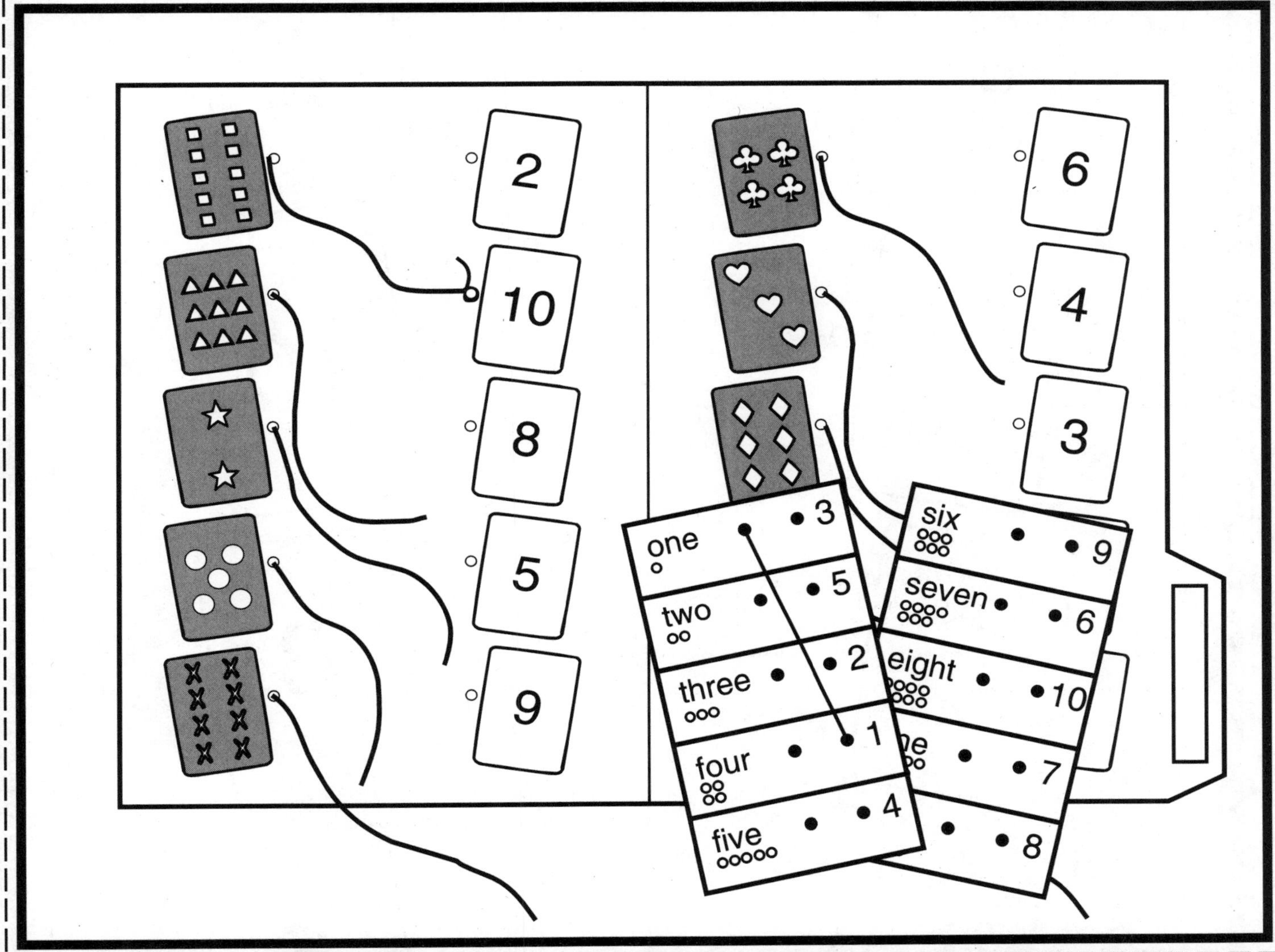

Card Mates

Matching the number of things to the symbol

Directions: Take out an activity card and the washable felt pen or erasable crayon. Set them aside. Open the folder. Find and name the matching number for each picture card. Wind the yarn over the brass fastener at the number. Pick up the activity card and put it in front of you. Draw a line with the pen to the number that matches the word. Name the number. For help, count the dots under the number word.

A book to read: <u>20,000 Baseball Cards Under the Sea</u>
by Jon Buller and Susan Schade

Card Mates

Card Mates

Card Mates

Activity Cards

one ○ •	• 3	six ○○○ ○○○ •	• 9
two ○○ •	• 5	seven ○○○○ ○○○ •	• 6
three ○○○ •	• 2	eight ○○○○ ○○○○ •	• 10
four ○○ ○○ •	• 1	nine ○○○ ○○○ ○○○ •	• 7
five ○○○○○ •	• 4	ten ○○○○○ ○○○○○ •	• 8

28. Coats

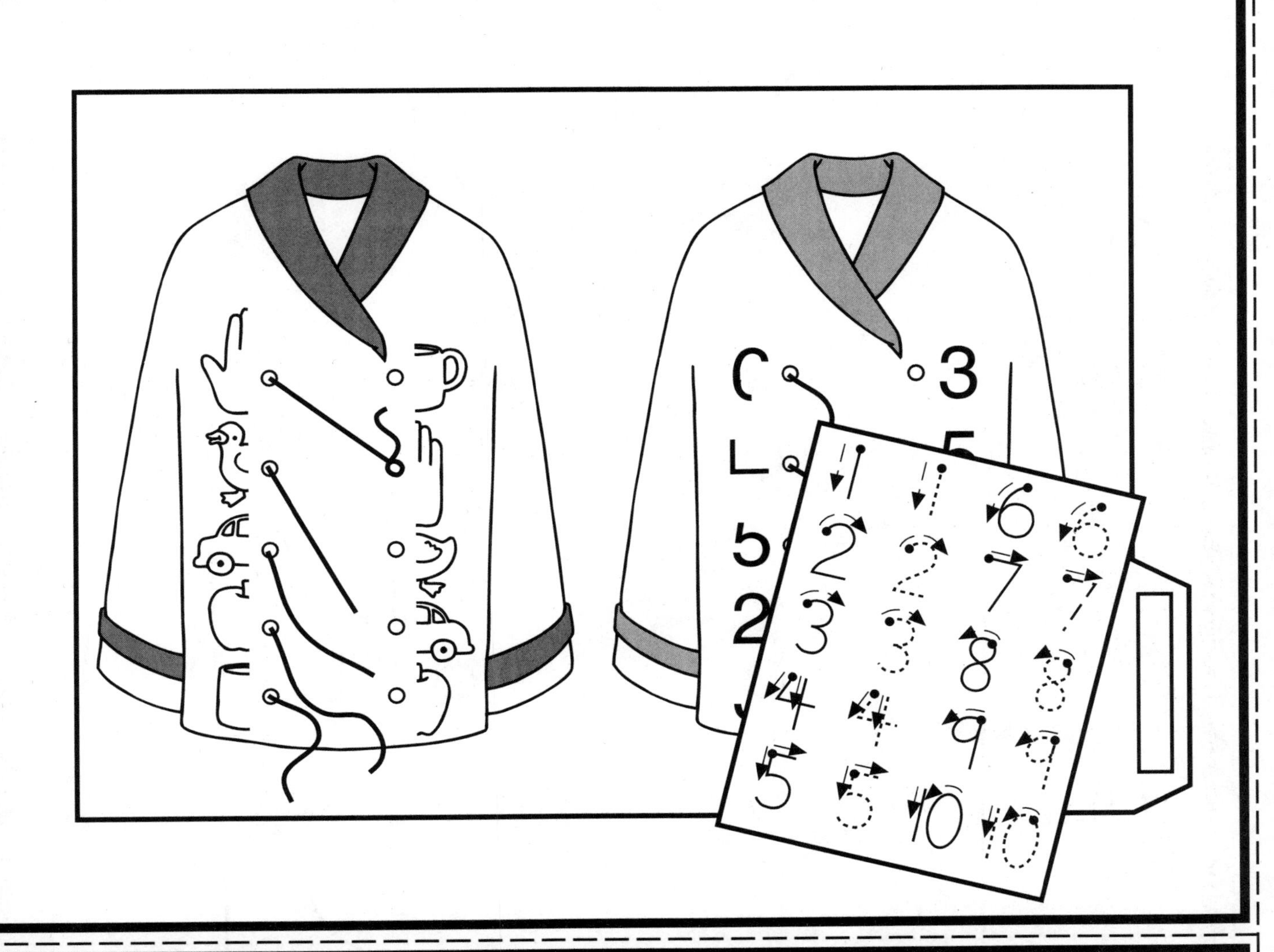

Coats

Object and number close-ups/number writing

Directions: Take out the activity card and the washable felt pen or erasable crayon. Set them aside. Open the folder. Look at the incomplete shapes or numbers on the left of the yarn-match coats. Find the matching shape part or number. Wind the yarn over the brass fastener at the answer. Pick up the activity card. Use the pen or crayon to trace the dotted lines for each number, watching the arrows. Say the name of each number.

A book to read: The Rag Coat by Lauren Mills

Coats

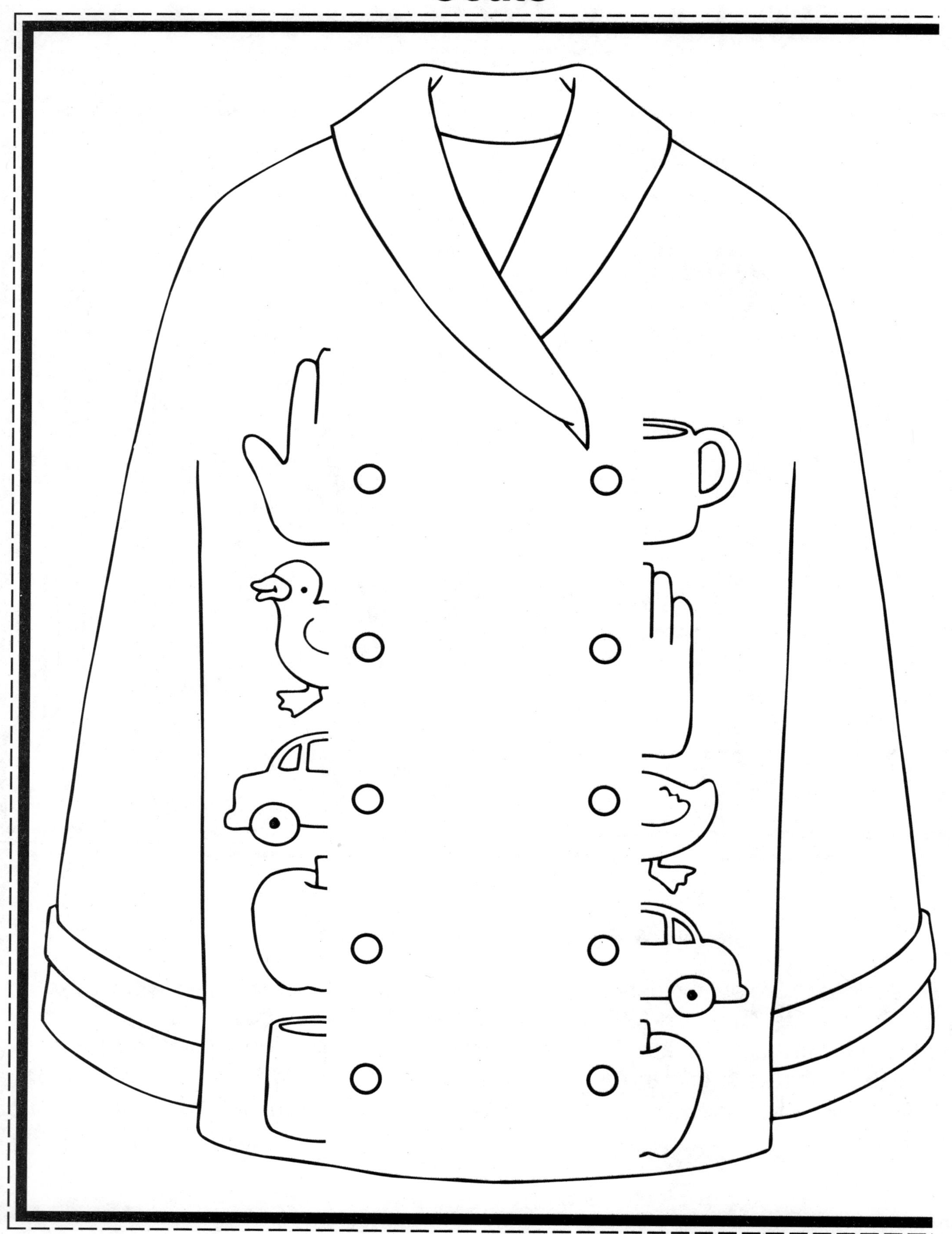

Coats

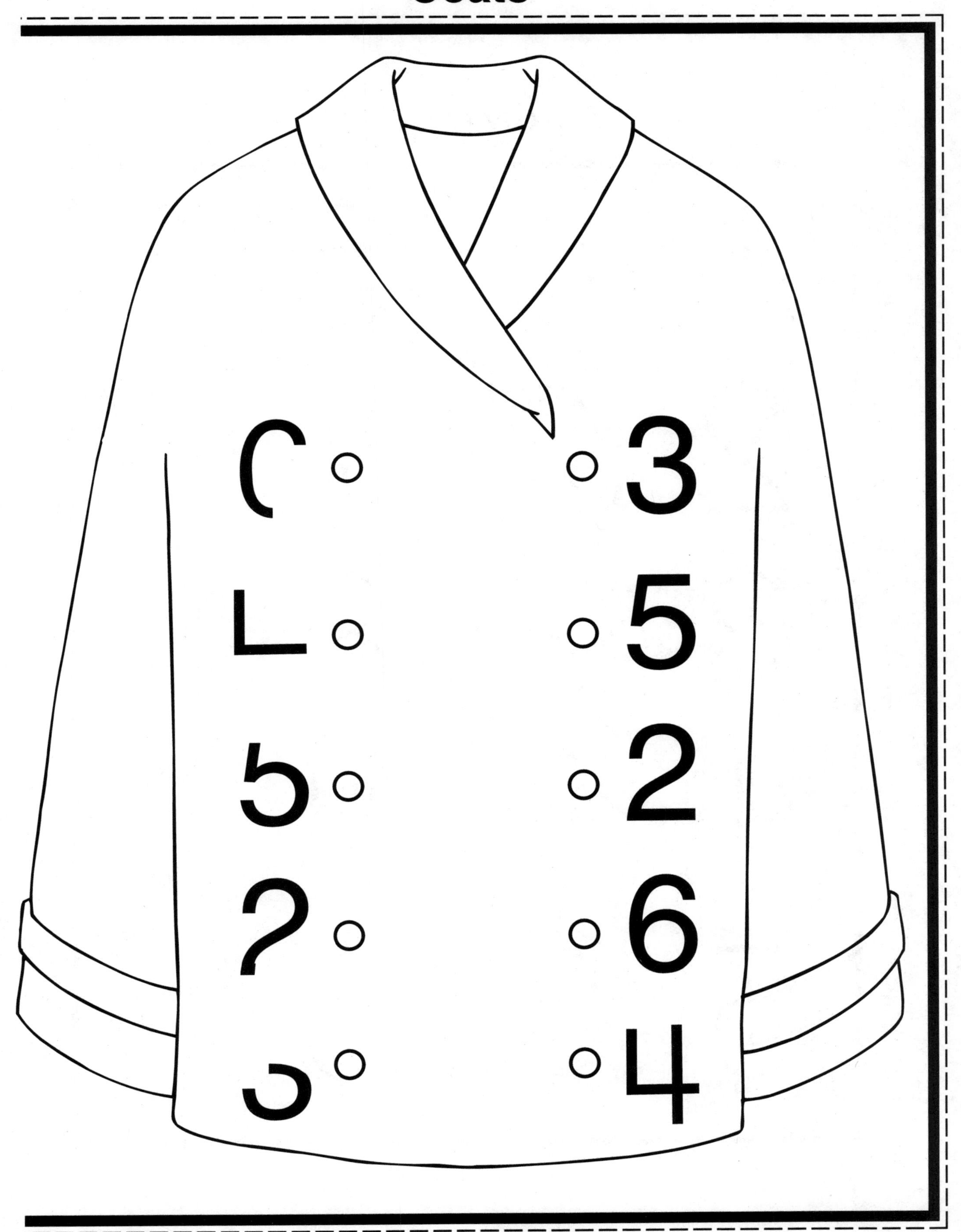

Coats

Activity Card

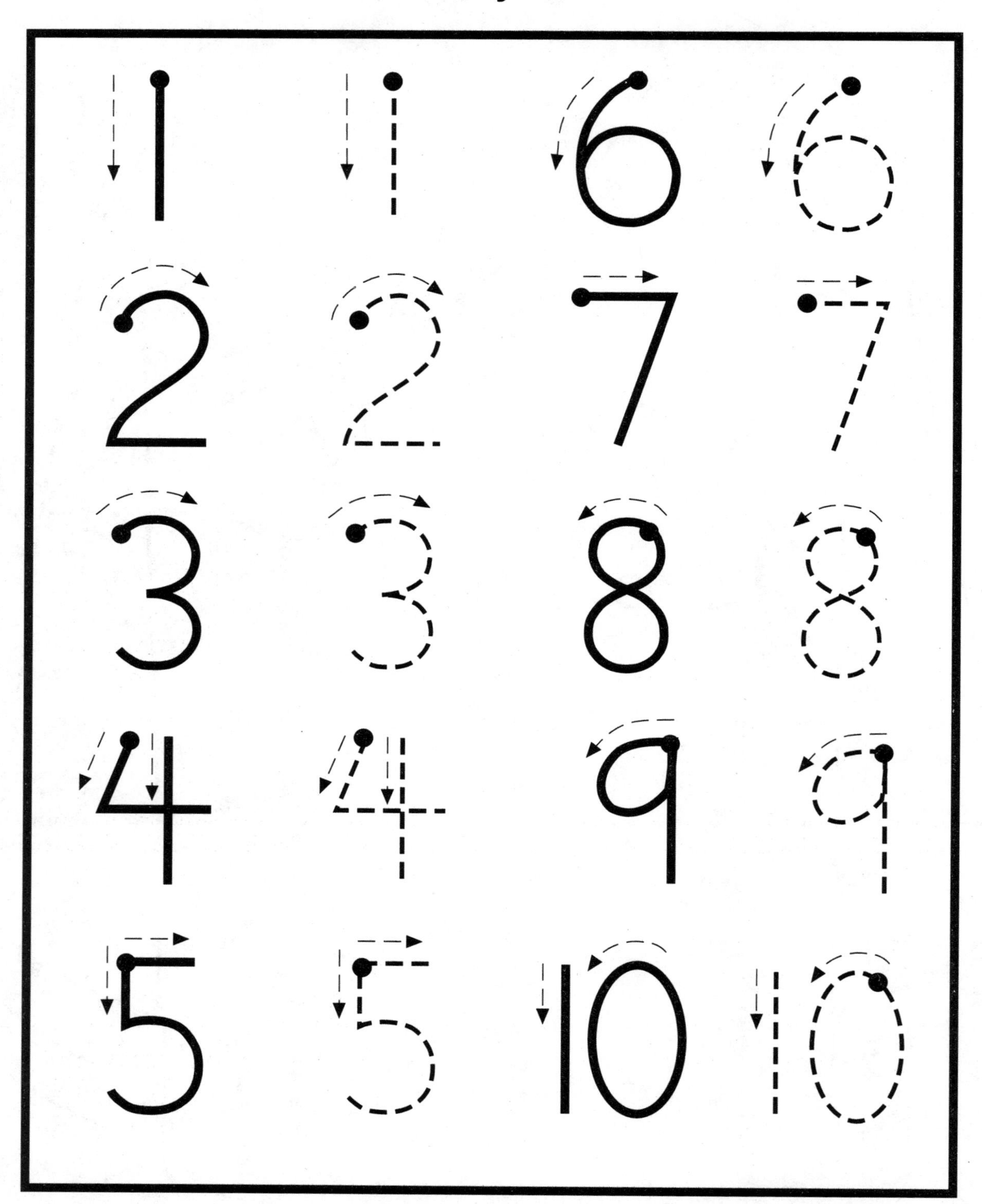

29. Dancing Stars

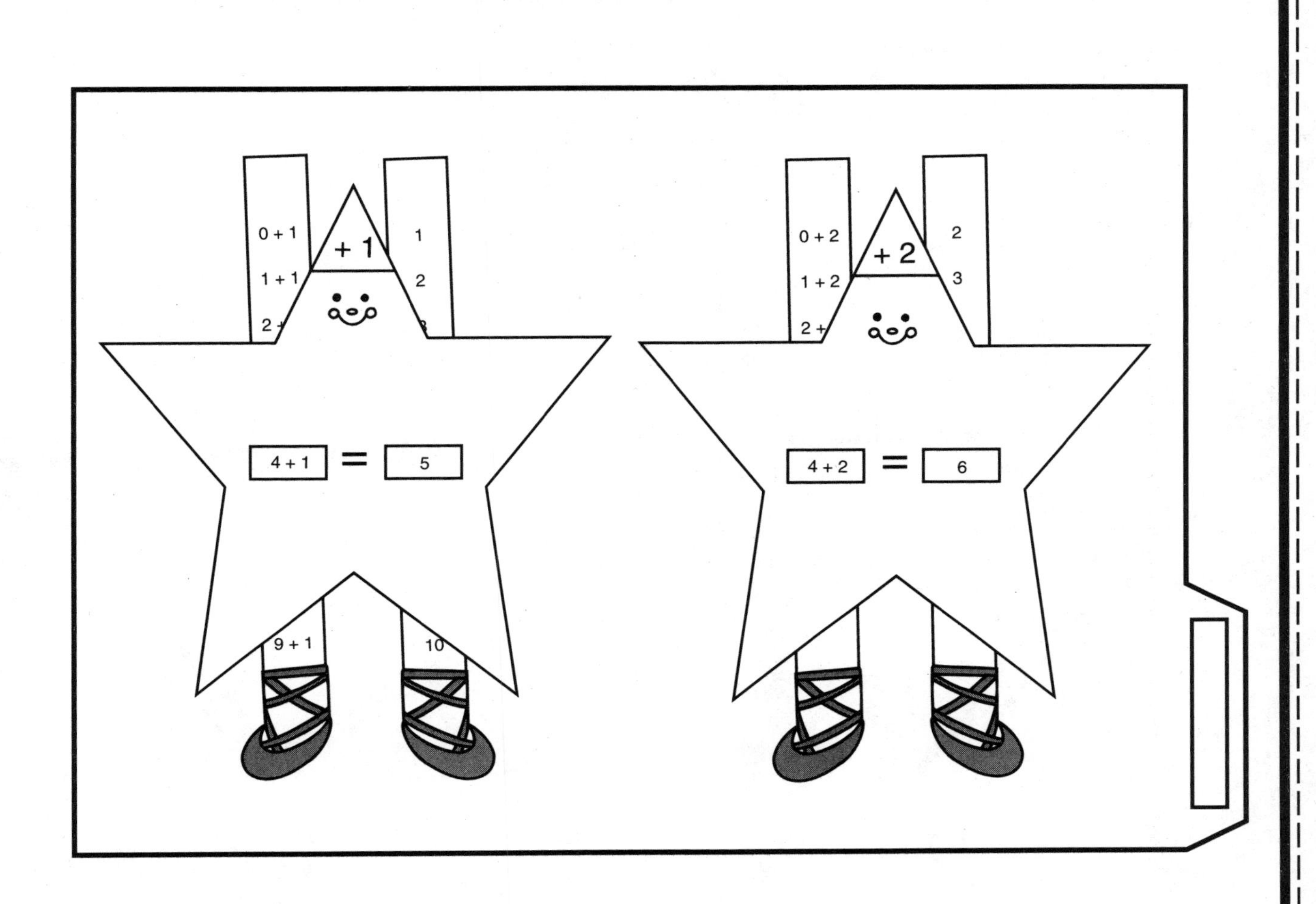

Dancing Stars

Adding 1 or 2 to a whole number

Directions: Open the folder. Slide the number fact and answer sliders in the stars. Match the number fact and the answer. You are adding 1 to the first number in the number sentence on the first star. You are adding 2 to the first number in the number sentence on the second star. Read each number sentence.

A book to read: The Dancing Star by Anne Rockwell

Dancing Stars

Dancing Stars

Dancing Stars

Number Fact Slider	Answer Slider	Number Fact Slider	Answer Slider
0 + 1	1	0 + 2	2
1 + 1	2	1 + 2	3
2 + 1	3	2 + 2	4
3 + 1	4	3 + 2	5
4 + 1	5	4 + 2	6
5 + 1	6	5 + 2	7
6 + 1	7	6 + 2	8
7 + 1	8	7 + 2	9
8 + 1	9	8 + 2	10
9 + 1	10		

30. Ships Ahoy!

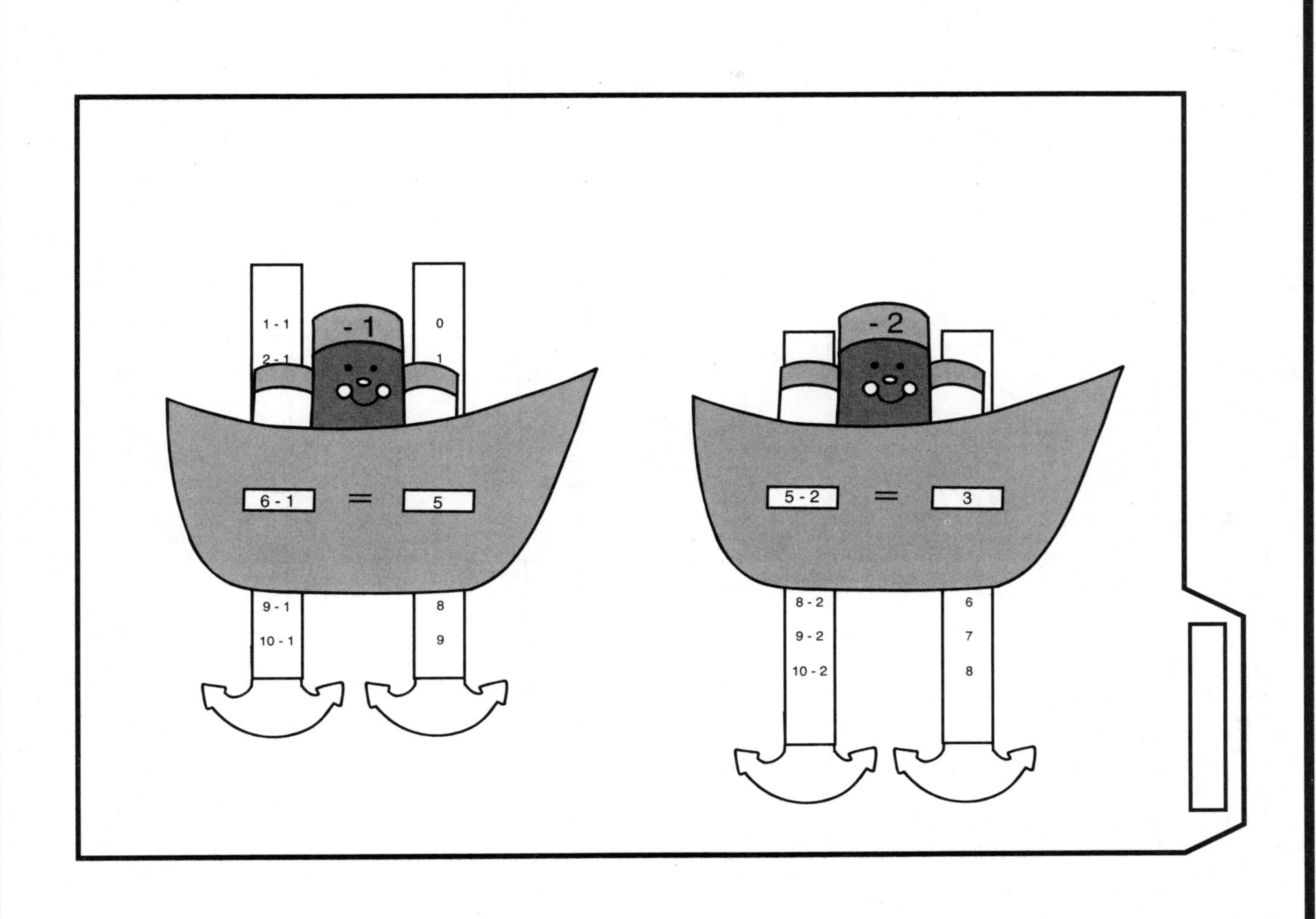

Ships Ahoy!

Taking away 1 or 2 from a whole number

Directions: Open the folder. Slide the number fact and answer sliders in the ships. Match the number fact and the answer. You are taking 1 away from the bigger number in the number sentence for the first ship. You are taking 2 away from the bigger number in the number sentence for the second ship. Read each number sentence.

A book to read: I Saw a Ship Sailing by Janina Domanska

Ships Ahoy!

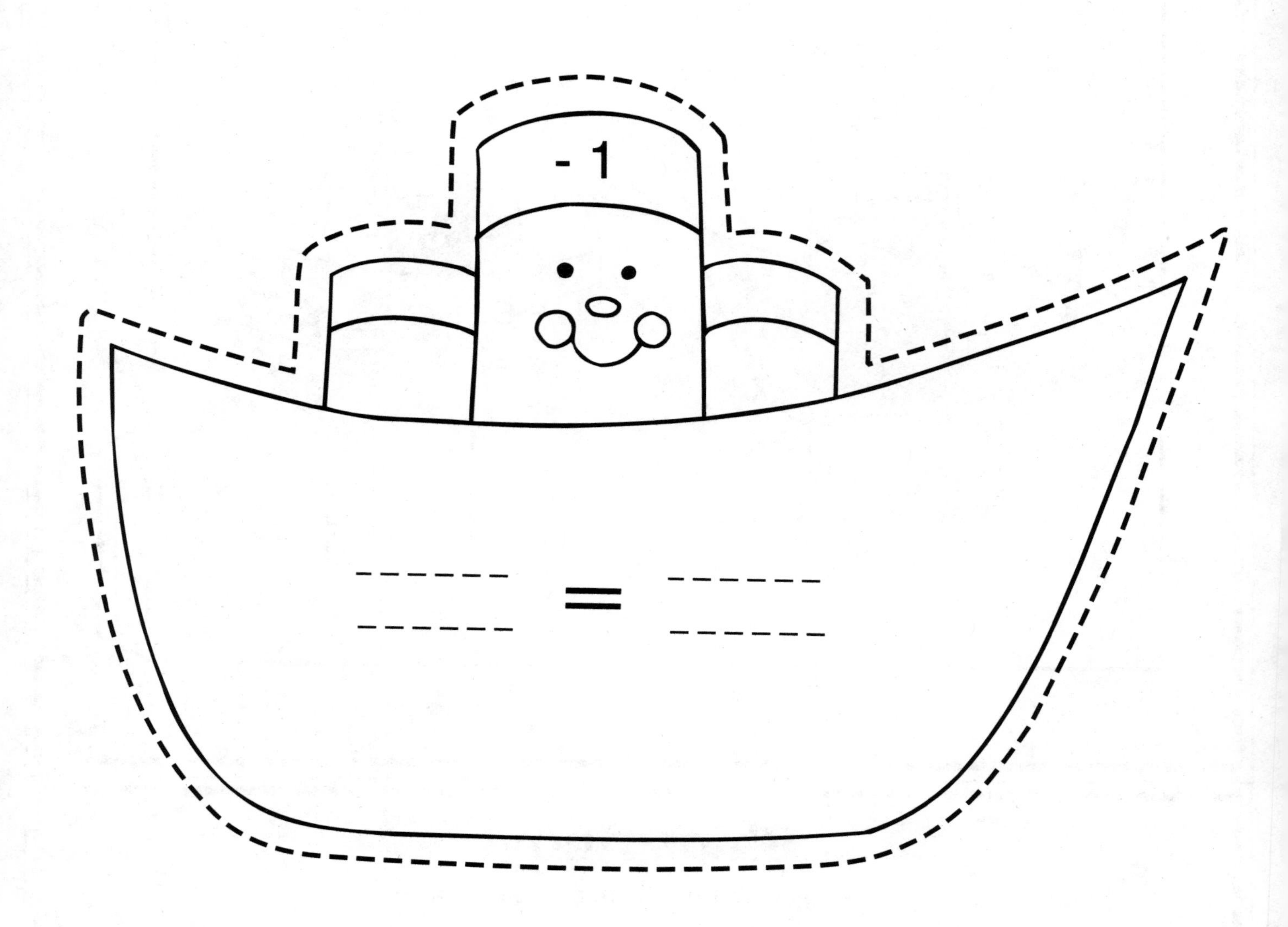

Ships Ahoy!

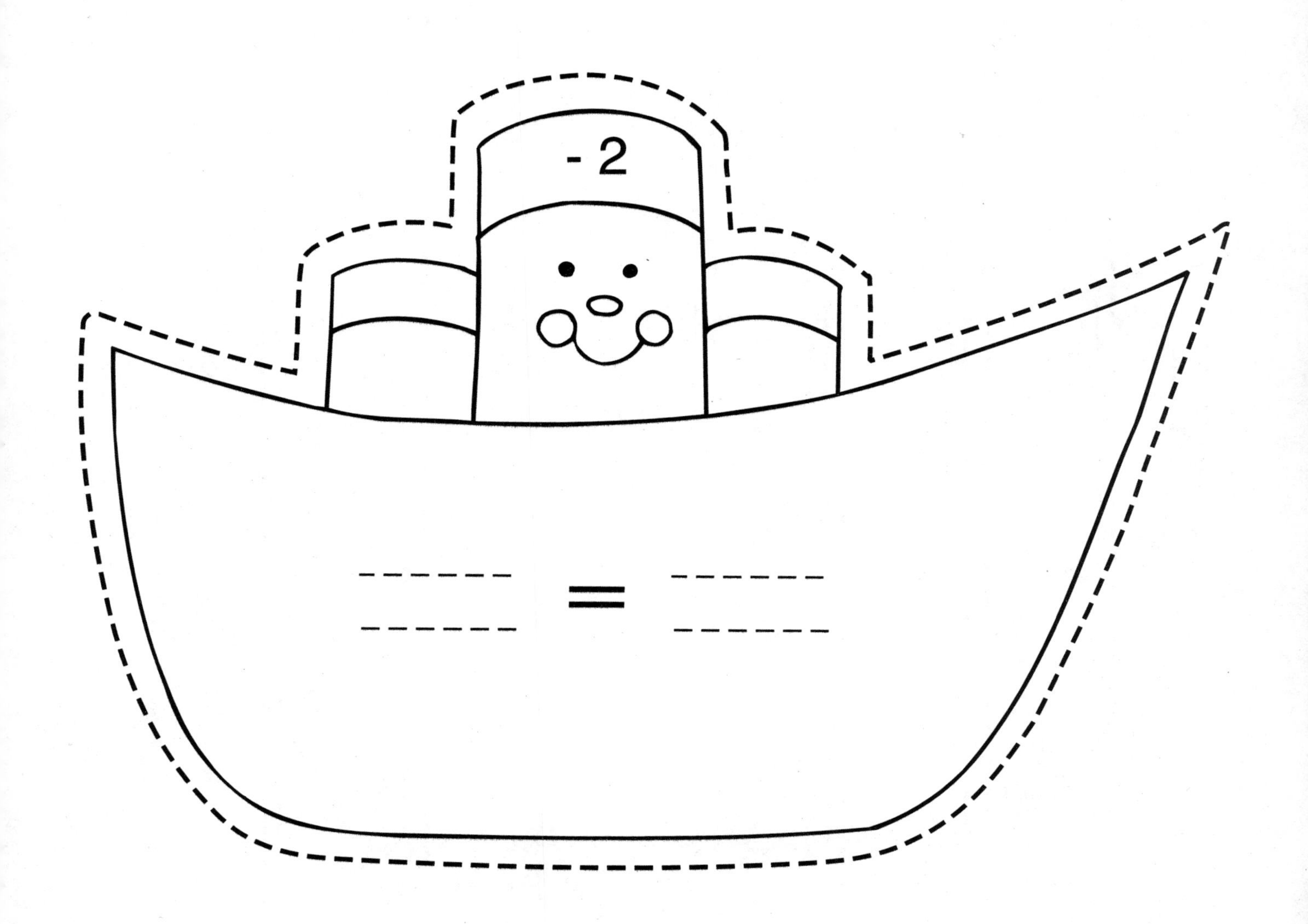

Ships Ahoy!

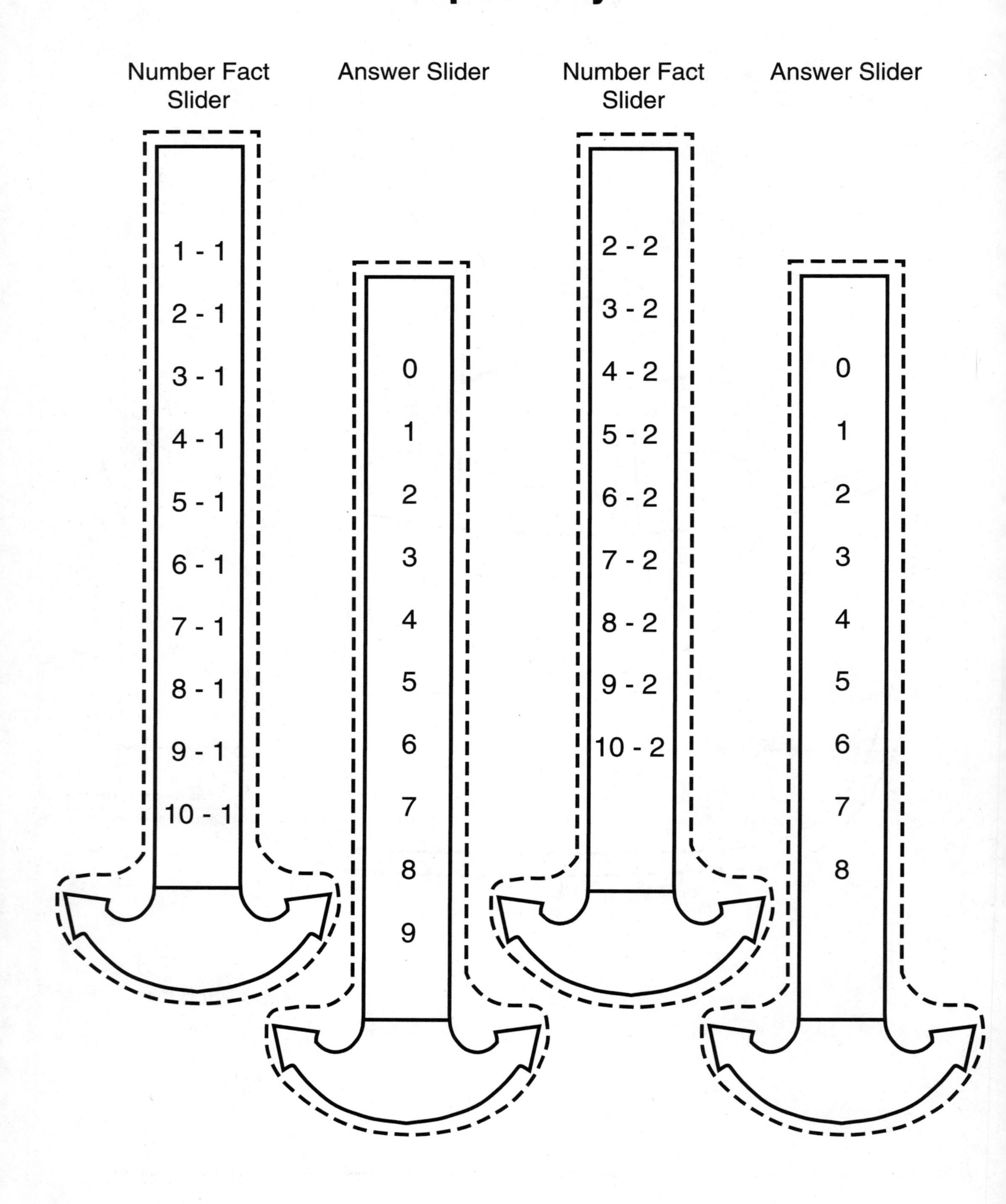
Number Fact Slider
1 - 1
2 - 1
3 - 1
4 - 1
5 - 1
6 - 1
7 - 1
8 - 1
9 - 1
10 - 1
Answer Slider
0
1
2
3
4
5
6
7
8
9
Number Fact Slider
2 - 2
3 - 2
4 - 2
5 - 2
6 - 2
7 - 2
8 - 2
9 - 2
10 - 2
Answer Slider
0
1
2
3
4
5
6
7
8

31. Spaceships

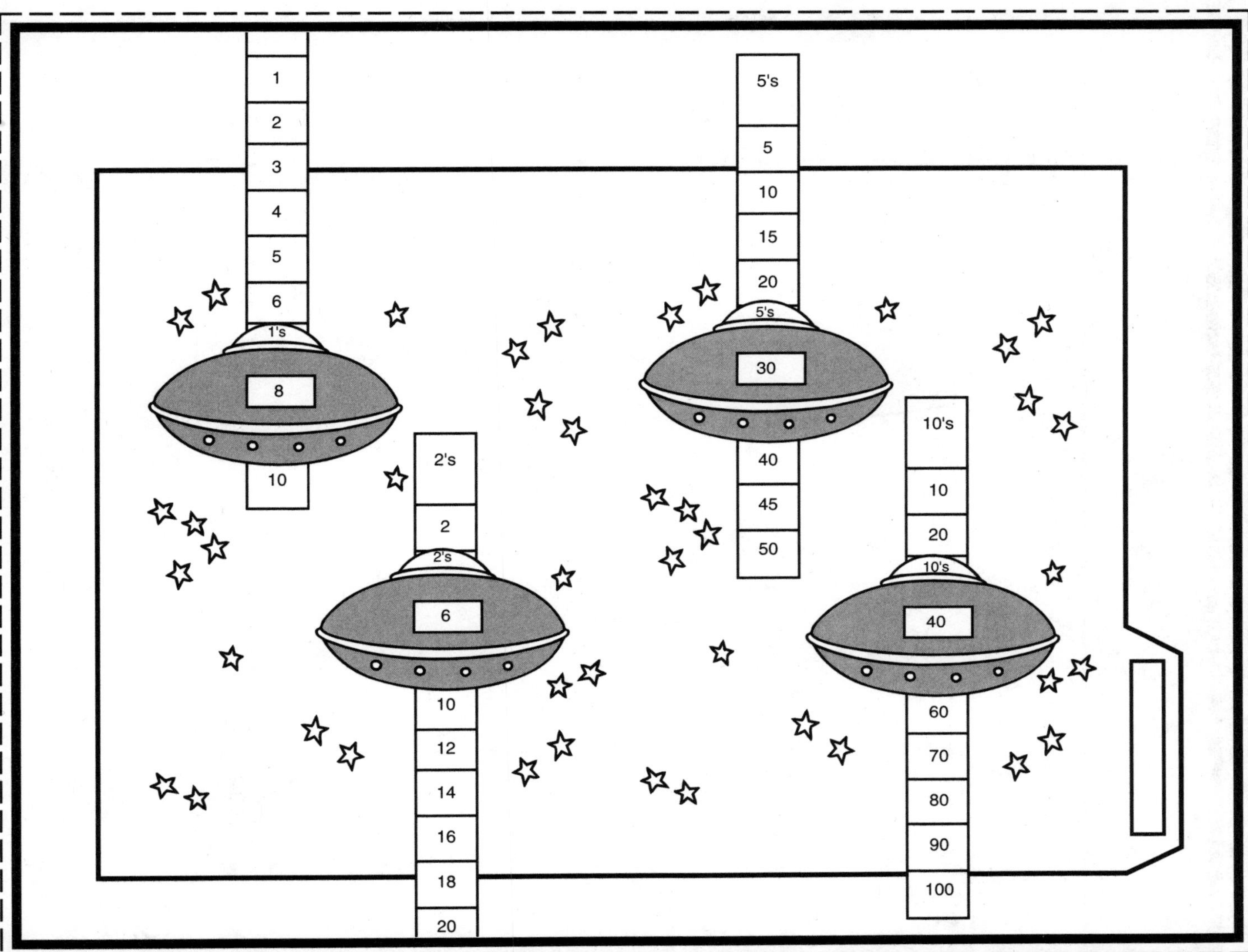

Spaceships

Counting by 1's, 2's, 5's, and 10's

Directions: Open the folder. Check the number at the top of each spaceship. Count either by 1's, 2's, 5's, or 10's. Pull the slider from the top and read each number as it shows in the window. Do it slowly first and faster when you know the counting.

A book to read: Professor Noah's Spaceship by Brian Wildsmith

Spaceships

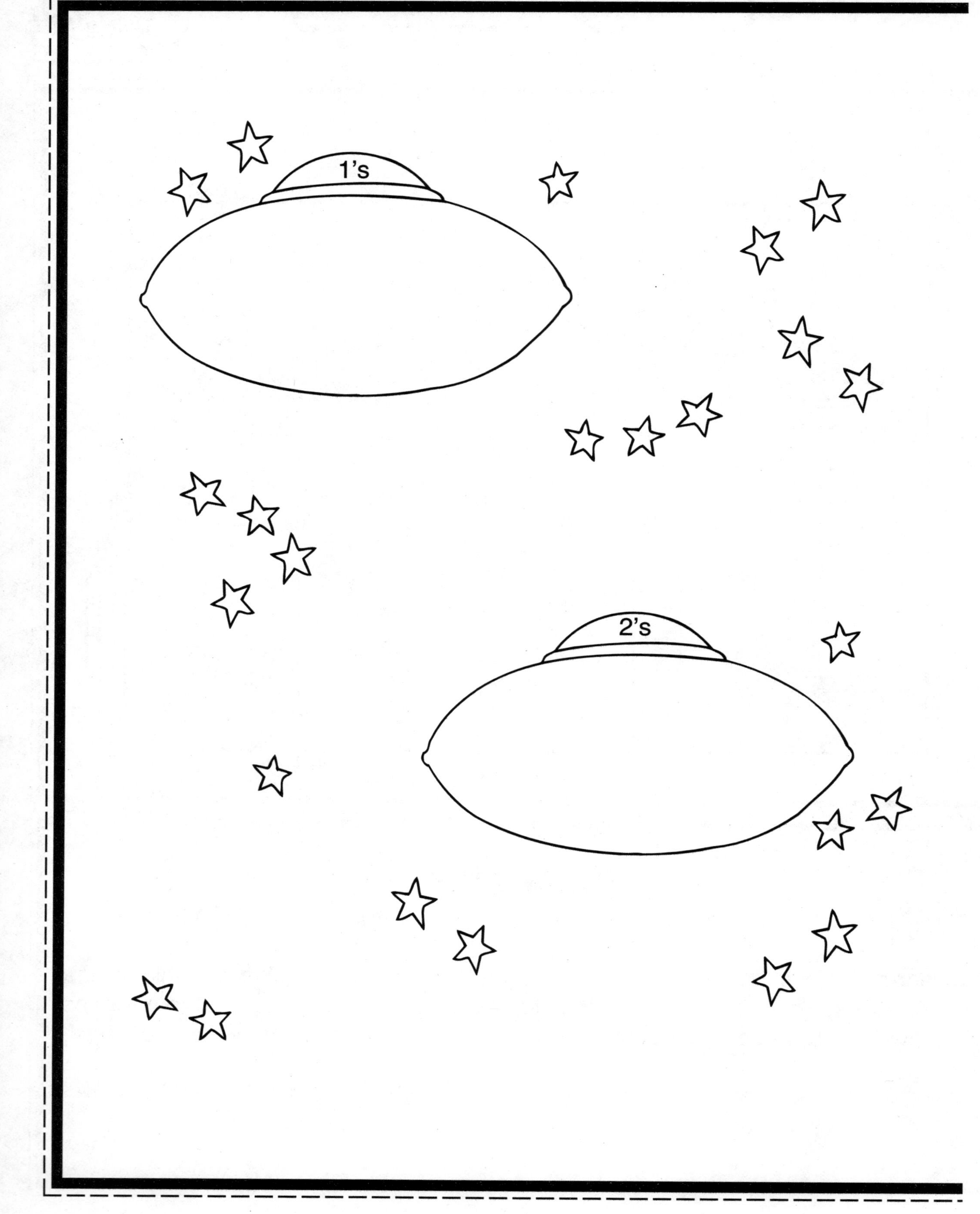

Spaceships

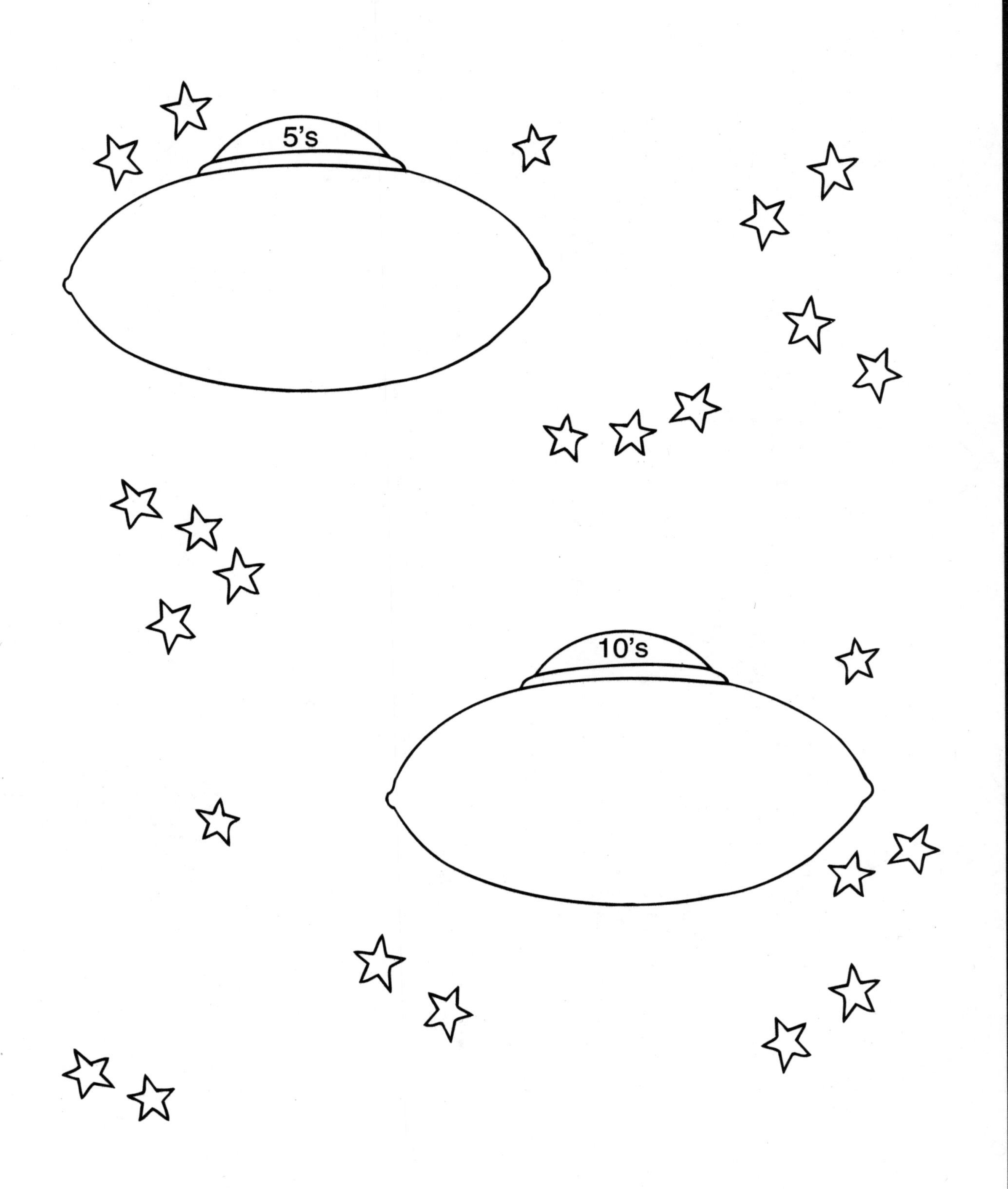

Spaceships

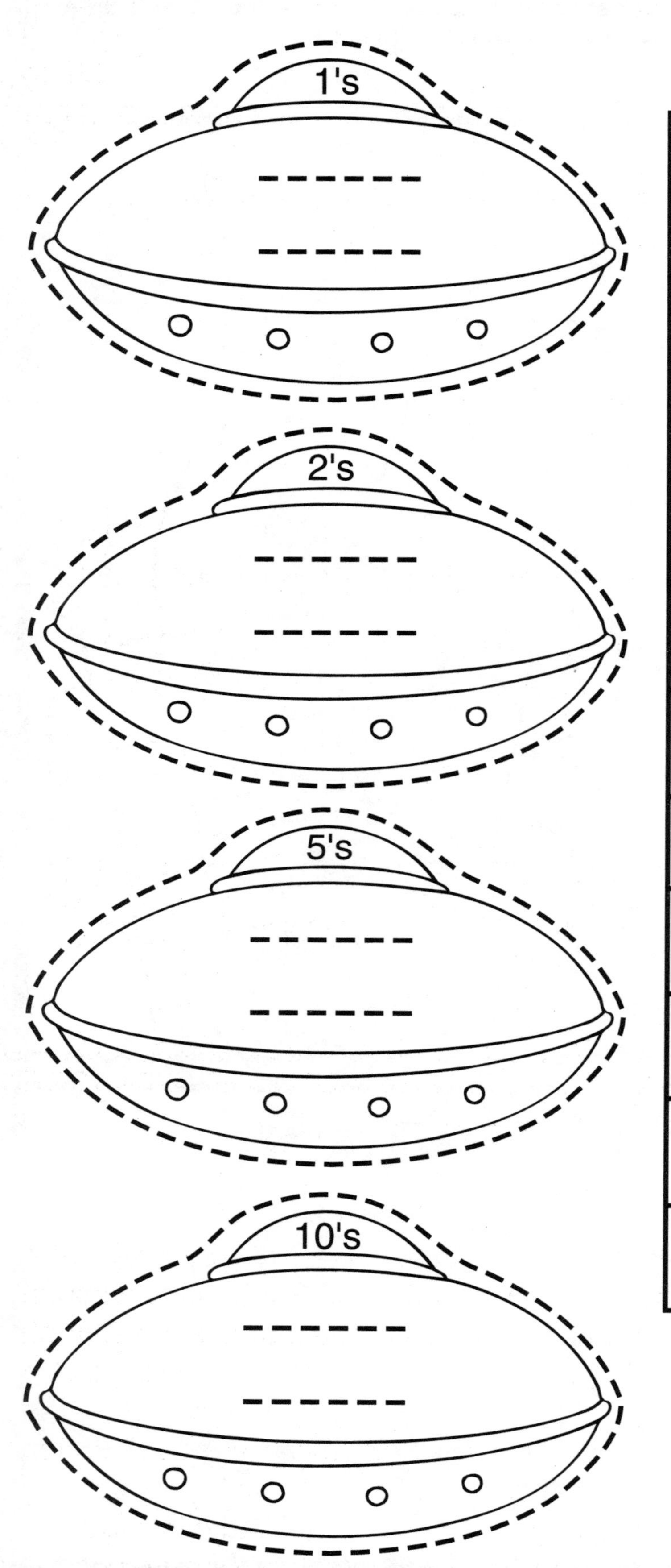

Cut apart on dotted lines.

1's	2's	5's	10's
1	2	5	10
2	4	10	20
3	6	15	30
4	8	20	40
5	10	25	50
6	12	30	60
7	14	35	70
8	16	40	80
9	18	45	90
10	20	50	100

32. Cat, Dog, and Mouse Race

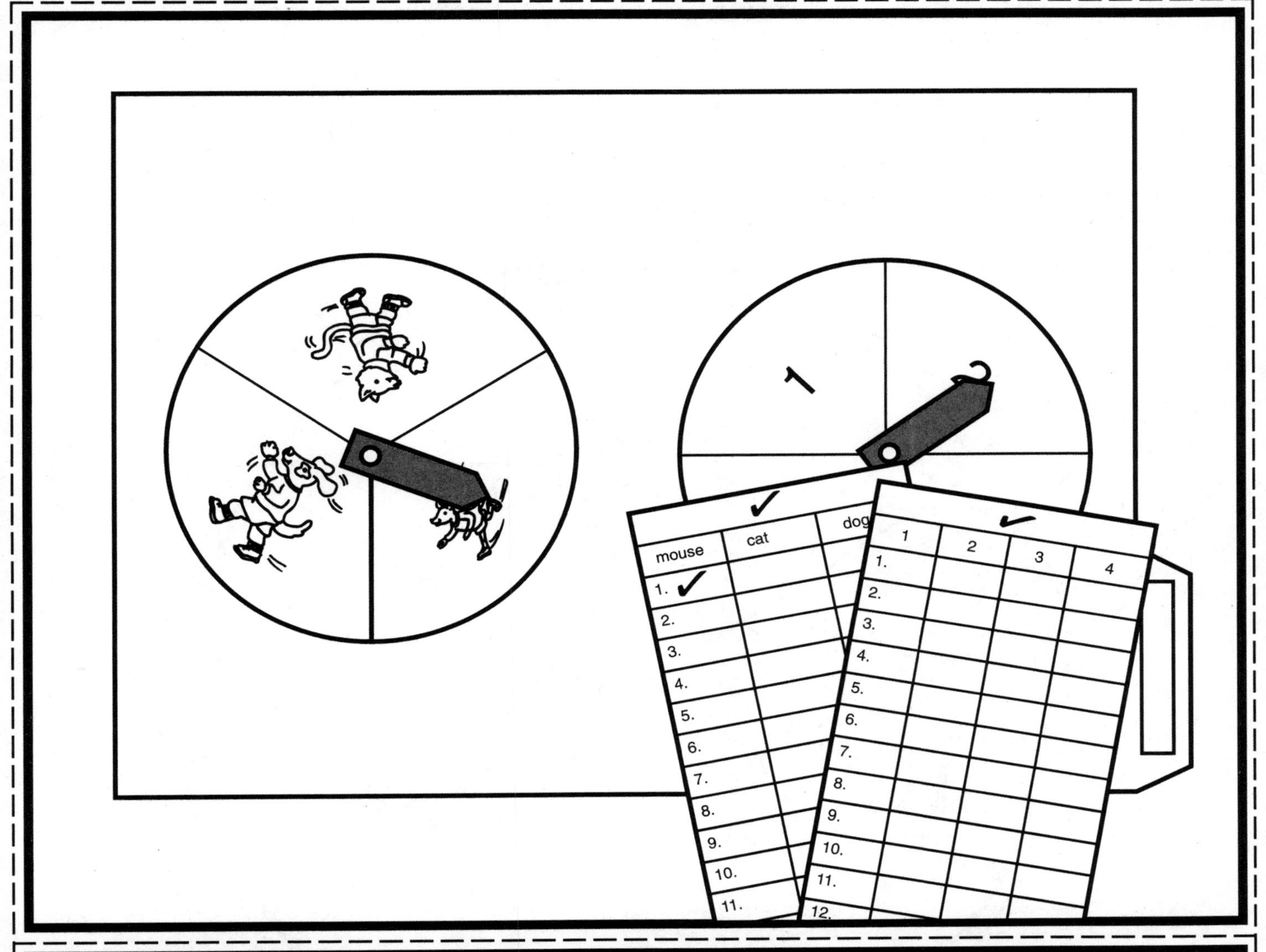

Cat, Dog, and Mouse Race

Beginning probability

Directions: Two children may play. Start with the cat, mouse, and dog spinner. Take a tally card for it and a felt pen or crayon. Spin the spinner in turn. See on which animal the spinner stops. Make a check in the column where it says the name of the animal. Keep taking turns spinning and marking. Which animal has the best chance of getting the most marks? Play ends when a player has filled in all the space for one of the animals. Play the same way with the number wheel and the number tally card. Which number has the best chance of getting the most marks?

A book to read: Here Comes That Cat! by Frank Asch and Vladimir Vagin

Cat, Dog, and Mouse Race

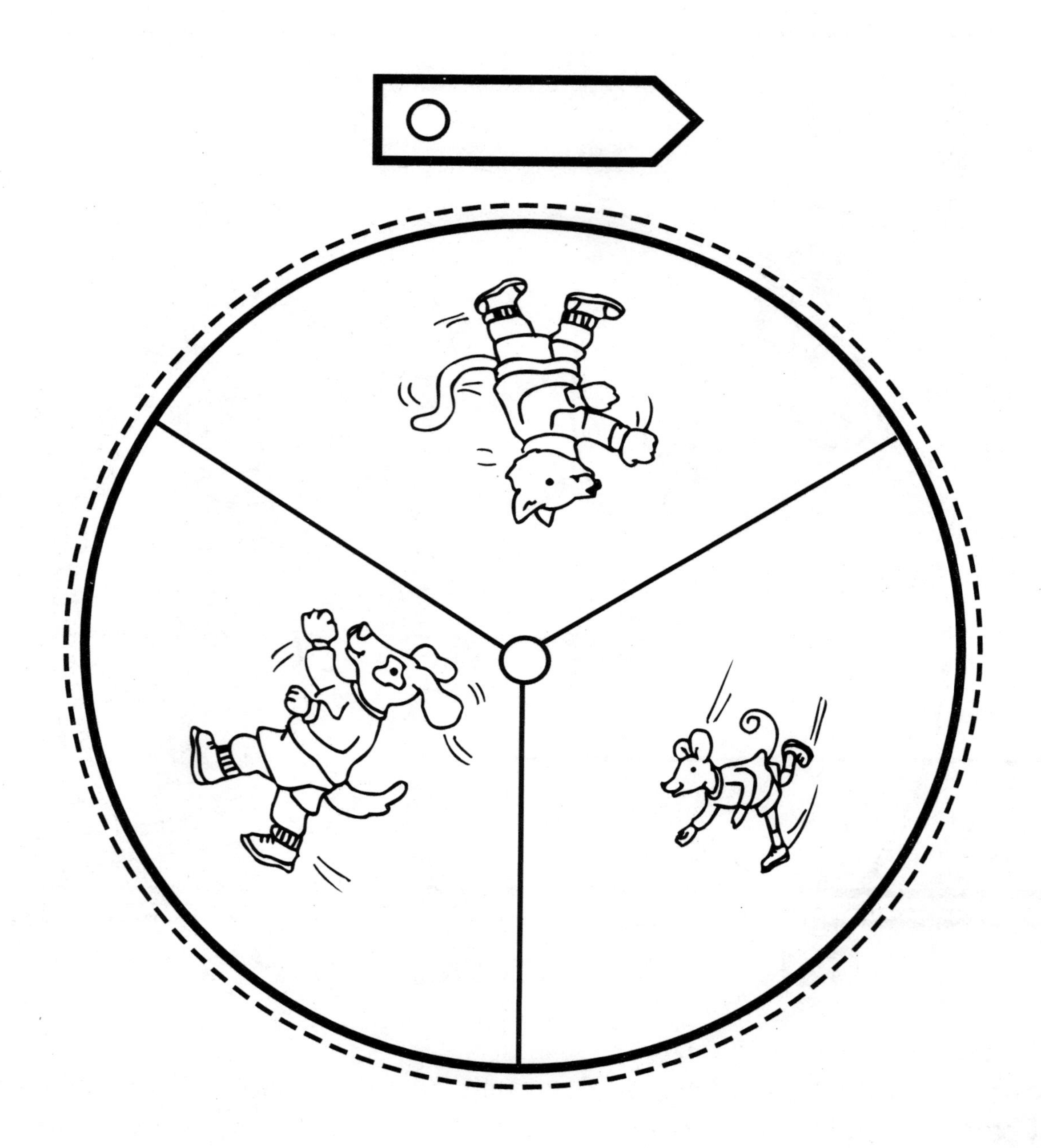

Cat, Dog, and Mouse Race

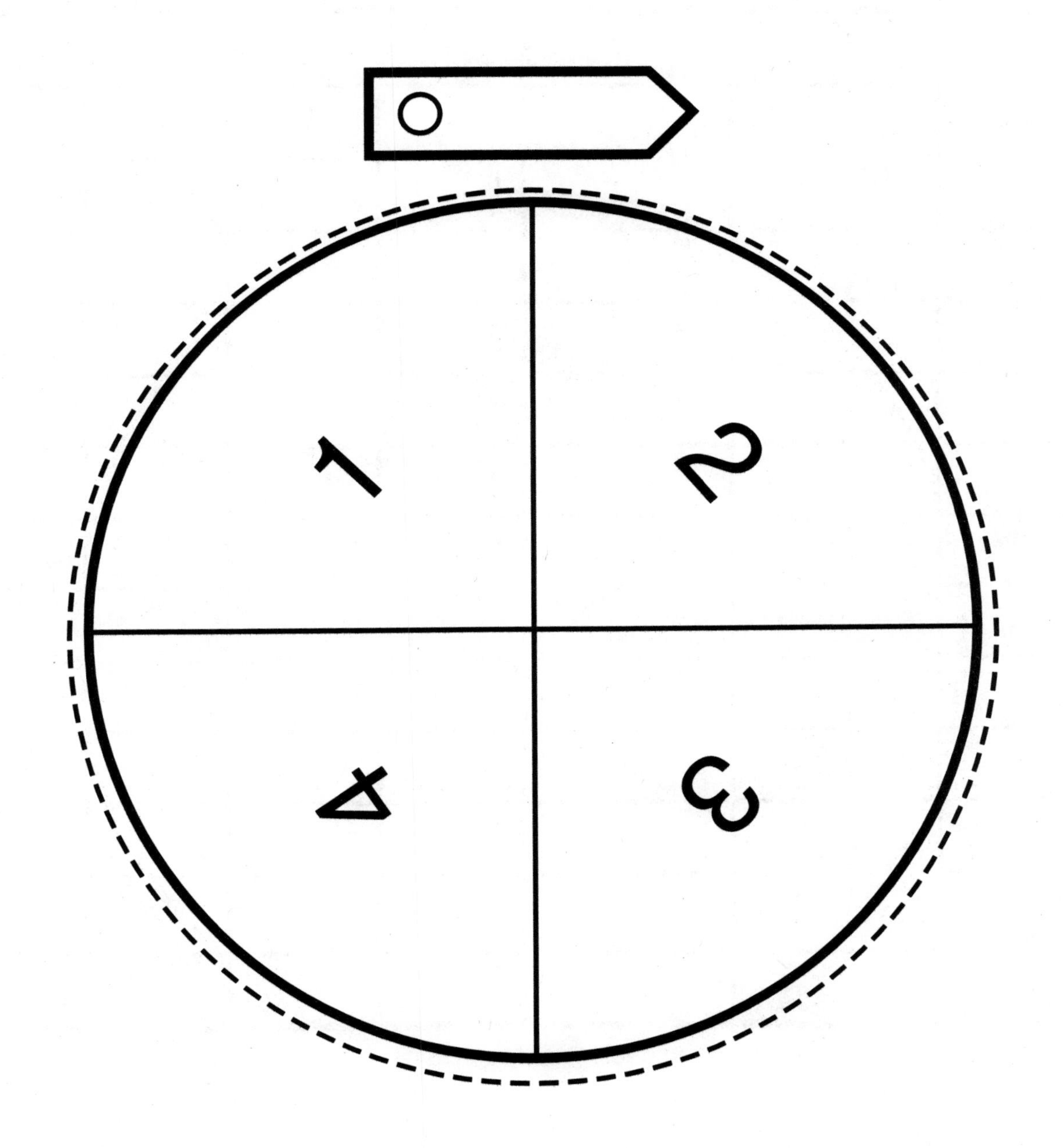

Cat, Dog, and Mouse Race

Tally Cards

Duplicate twice.

✔			✔			
mouse	cat	dog	1	2	3	4
1.			1.			
2.			2.			
3.			3.			
4.			4.			
5.			5.			
6.			6.			
7.			7.			
8.			8.			
9.			9.			
10.			10.			
11.			11.			
12.			12.			

33. Pies

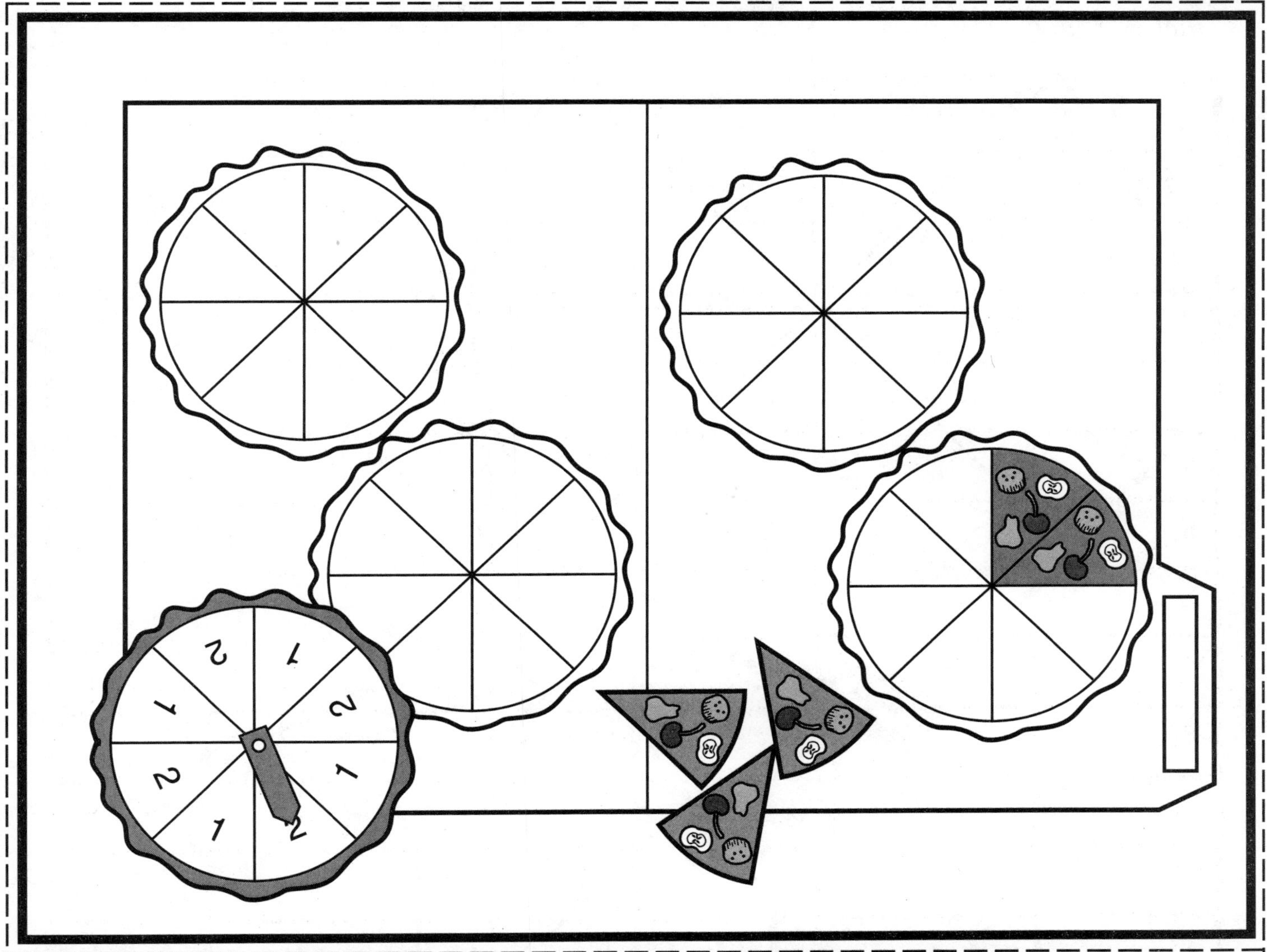

Pies

Beginning fractions

Directions: Two children may play. Take out the pie pieces and the pie spinner. Put the pie pieces in a pile. Open the folder. Choose two pies to play with. In turn, spin the spinner and check the number it stops on. Take as many pieces of pie as that number. Put the pieces on the pie. See who can make a whole pie first.

A book to read: Crawly Bug and Firehouse Pig by Toby Speed

Pies

Duplicate twice.

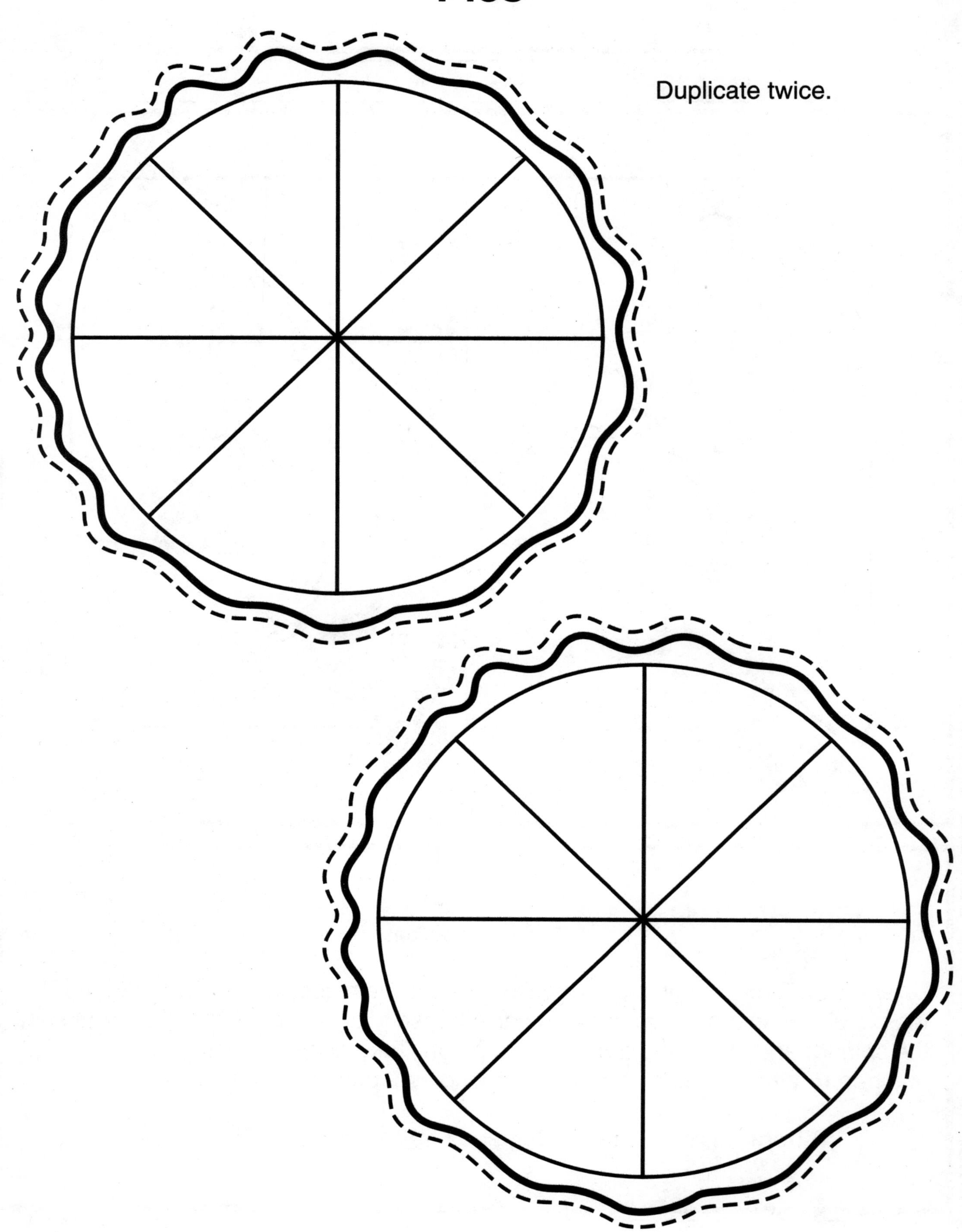

Pies

Duplicate twice.
Cut apart.

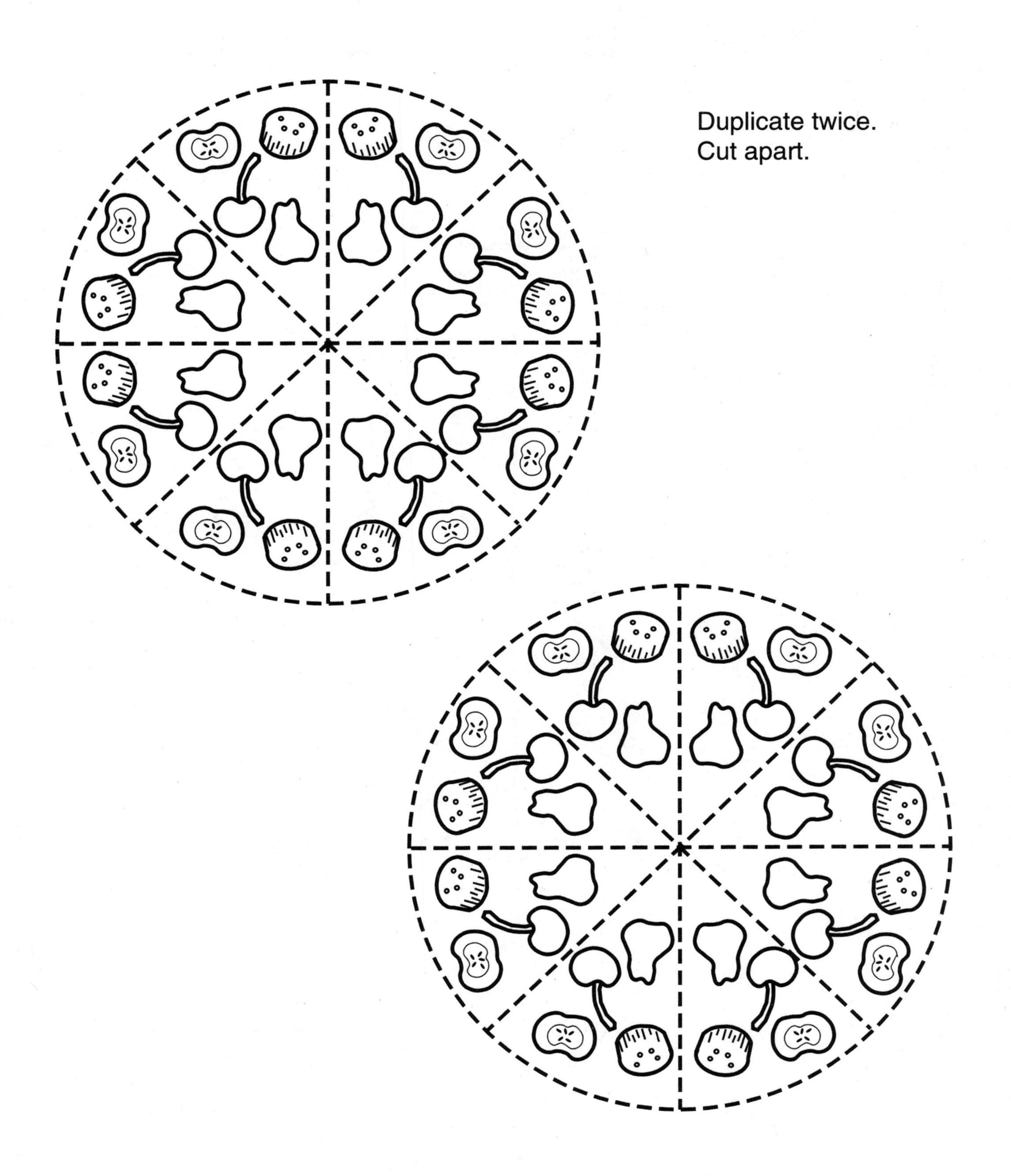

Pies

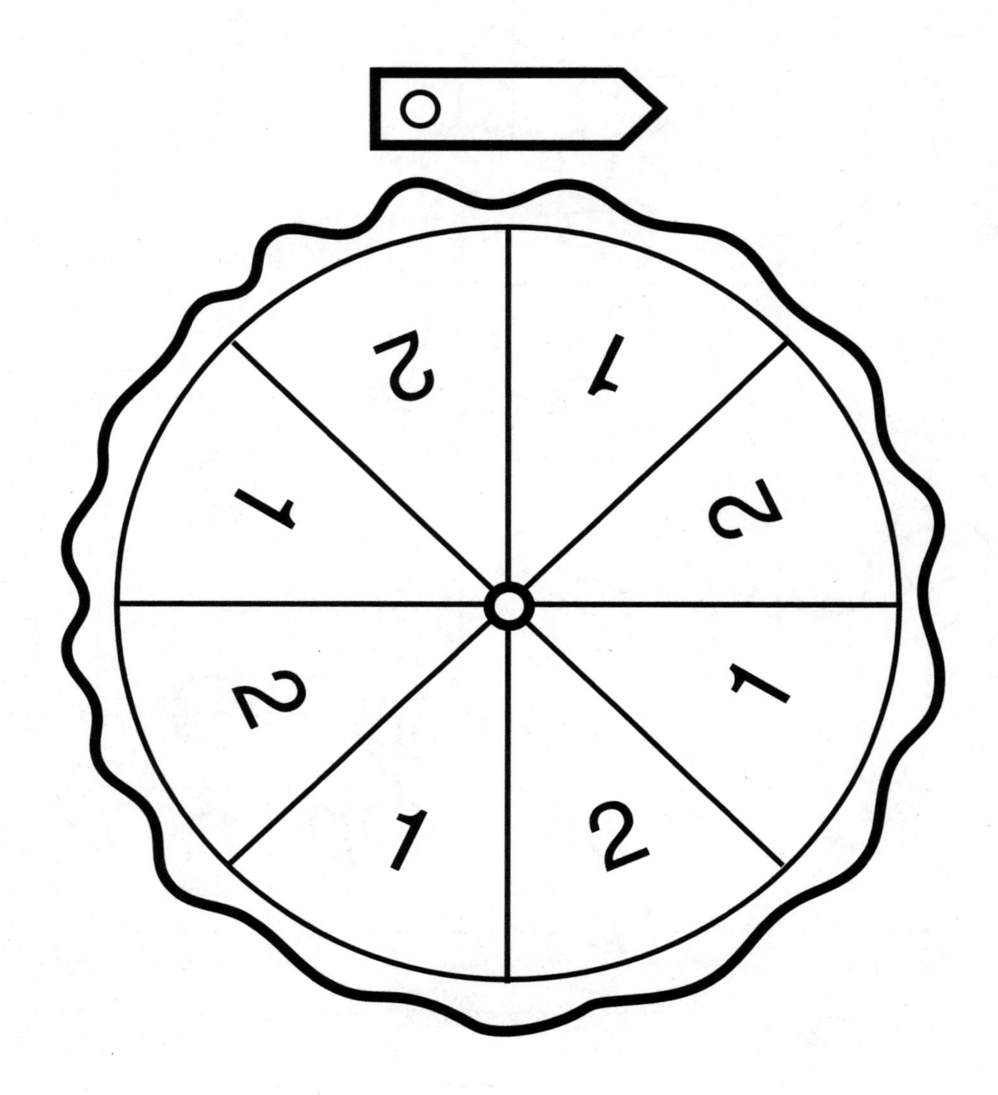

34. Mouse Time

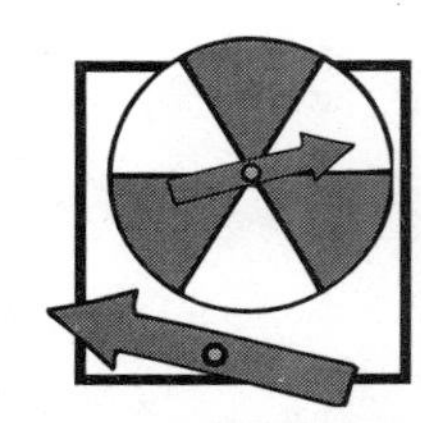

Mouse Time

Telling time

Directions: Two children may play. Place the mouse cards in a pile on the playing area. Choose one mouse time card to play with. Spin the spinner on the big clock in turn. Check the number that the spinner stops closest to. That is your hour. Take a mouse and put it on the matching clock on your mouse time card. Name the hour. The player who fills all the mouse times on his or her card first gets the cheese!

A book to read: Hickory, Dickory, Dock by Robin Miller and Suzanne Durancean

Mouse Time

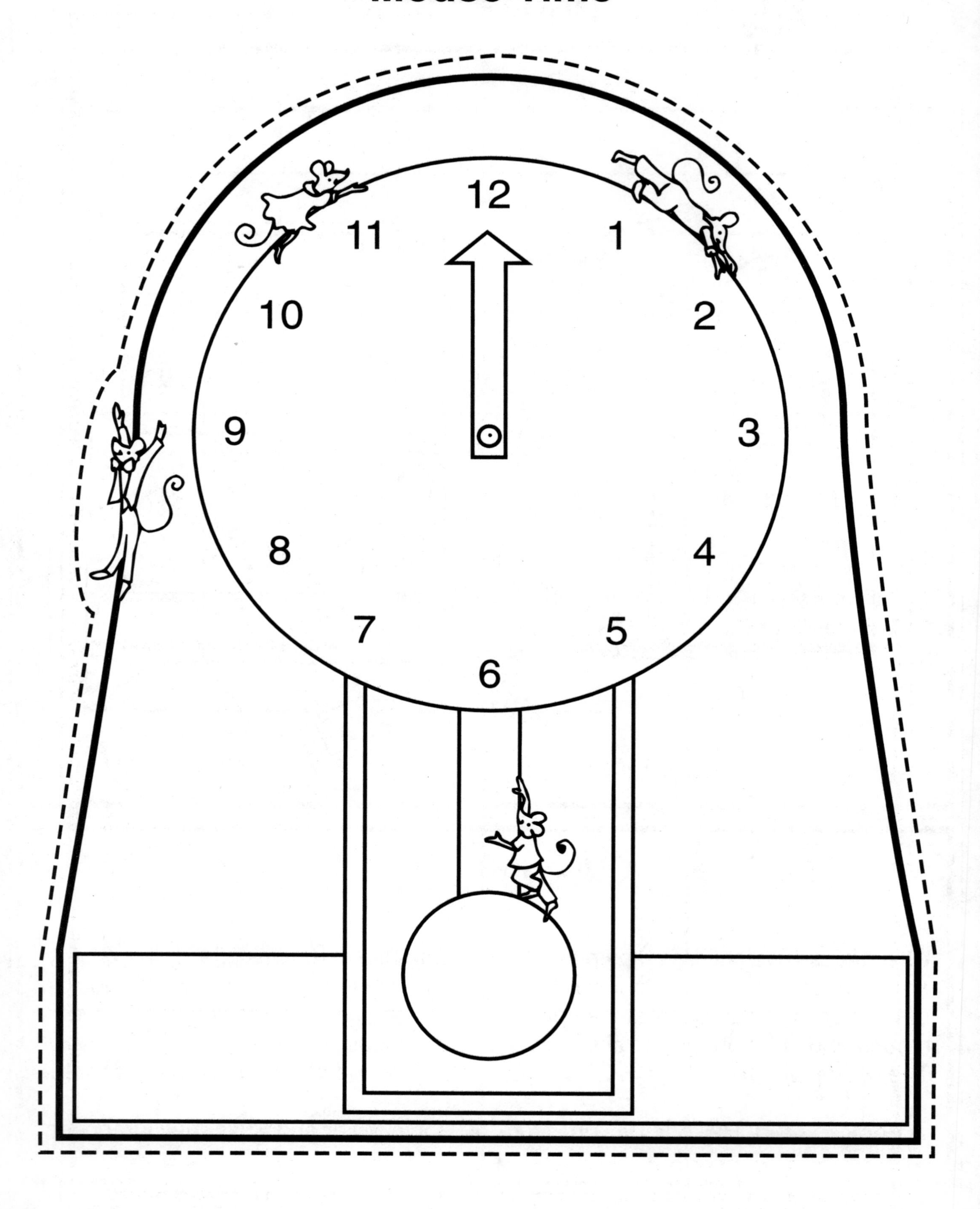

Mouse Time

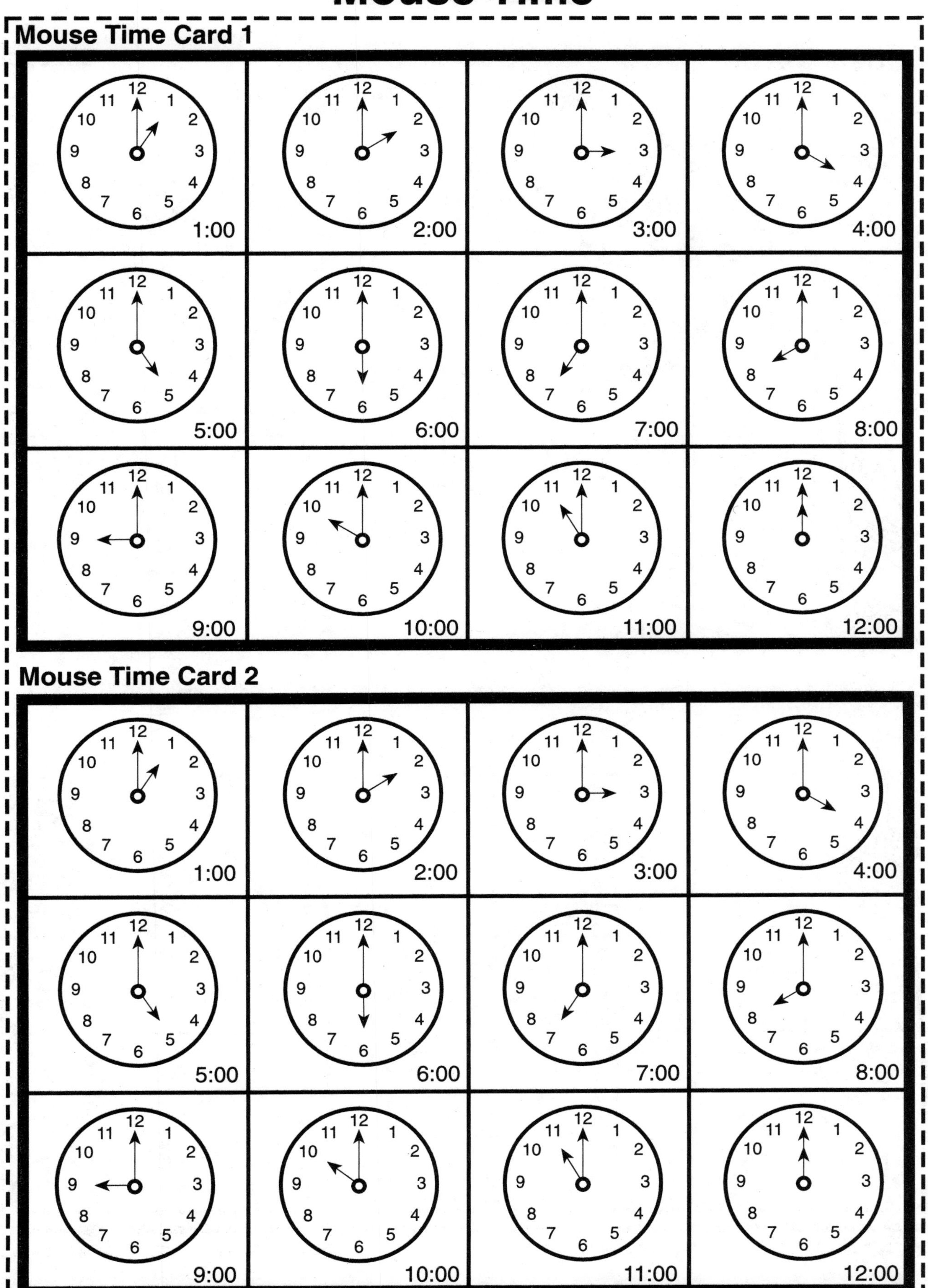
Mouse Time Card 1
1:00
2:00
3:00
4:00
5:00
6:00
7:00
8:00
9:00
10:00
11:00
12:00
Mouse Time Card 2
1:00
2:00
3:00
4:00
5:00
6:00
7:00
8:00
9:00
10:00
11:00
12:00

Mouse Time

Mouse Cards

Cut apart.